# 21st Century Communication 3

## LISTENING, SPEAKING, AND CRITICAL THINKING

**Second Edition**

LYNN BONESTEEL
RICHARD WALKER

Australia · Brazil · Canada · Mexico · Singapore · United Kingdom · United States

National Geographic Learning,
a Cengage Company

***21st Century Communication 3*, Second Edition**
**Lynn Bonesteel and Richard Walker**

Publisher: Andrew Robinson
Executive Editor: Sean Bermingham
Senior Development Editor: Melissa Pang
Development Editors: Sophia Khan, Rayne Ngoi
Assistant Editor: Dawne Law
Director of Global Marketing: Ian Martin
Heads of Regional Marketing:
Charlotte Ellis (Europe, Middle East and Africa)
Justin Kaley (Asia and Greater China)
Irina Pereyra (Latin America)
Joy MacFarland (US and Canada)
Product Marketing Manager: Tracy Bailie
Senior Production Controller: Tan Jin Hock
Senior Media Researcher: Leila Hishmeh
Senior Designer: Heather Marshall
Operations Support: Hayley Chwazik-Gee
Manufacturing Buyer: Terrence Isabella
Composition: MPS North America LLC



Student's Book with Spark platform access:
ISBN-13: 978-0-357-85599-7

Student's Book:
ISBN-13: 978-0-357-86198-1

**National Geographic Learning**
200 Pier 4 Boulevard
Boston, MA 02210
USA

Locate your local office at **international.cengage.com/region**

Visit National Geographic Learning online at **ELTNGL.com**
Visit our corporate website at **www.cengage.com**

Printed in Singapore
Print Number: 01 Print Year: 2023

# Topics and Featured Speakers

## 1

### Rethinking Success

Q: How do we define success?

**JIA JIANG**
*What I Learned From 100 Days of Rejection*

## 2

### Changemakers

Q: How can we make a difference in the world?

**MELATI WIJSEN**
*A Roadmap for Young Changemakers*

## 3

### Say It Your Way

Q: How can we become better communicators?

**ERIN McKEAN**
*Go Ahead, Make Up New Words*

## 4

### Stress: Friend or Foe?

Q: How do our attitudes affect us?

**KELLY McGONIGAL**
*Why Stress is Good for You*

## 5

### A Helping Hand

Q: When does helping really help?

**ASHA DE VOS**
*A Hero On Every Coastline*

## 6

### Be Your Own Boss

Q: Should we work for ourselves?

**BEL PESCE**
*5 Ways to Kill Your Dreams*

## 7

### Live Long, Live Well

Q: What steps can we take to live healthier lives?

**MATT WALKER**
*Sleep is Your Superpower*

## 8

### Beyond Limits

Q: How do we define our limits?

**PHIL HANSEN**
*Embrace the Shake*

# Scope and Sequence

| | LESSONS | | | |
|---|---|---|---|---|
| D COMMUNICATION | E VOCABULARY | F VIEWING | G CRITICAL THINKING | H PRESENTATION |
| A talk about how to define success<br>Create a definition of success, using a story to illustrate | Words related to opportunities and obstacles<br>Phrasal verbs with *turn* | TED Talk *What I Learned from 100 Days of Rejection* **Jia Jiang**<br>Pronunciation Skill Intonation in lists | Analyze an infographic comparing the goals of two different generations<br>Synthesize and evaluate ideas about success and failure | **Individual presentation:** Two successful people<br>Presentation Skill Use humor to connect with an audience |
| A conversation about making a difference in the local community<br>Create a proposal to help the local community | Words related to equality<br>The suffix *-or* | TED Talk *A Roadmap for Young Changemakers* **Melati Wijsen**<br>Pronunciation Skill Contrasting intonation | Analyze an infographic about youth beliefs and behavior<br>Synthesize and evaluate ideas about making a difference | **Individual presentation:** Three individuals or organizations that make a difference<br>Presentation Skill Show enthusiasm |
| An advertisement for a competition<br>Create and explain new blend words | Words related to language<br>Collocations with *attention* | TED Talk *Go Ahead, Make Up New Words* **Erin McKean**<br>Pronunciation Skill Stress in compound words | Analyze an infographic about how a new word enters the dictionary<br>Synthesize and evaluate ideas about the acceptance of new words | **Group presentation:** Three uncommon but useful words<br>Presentation Skill Encourage audience participation |
| A conversation between two students about their problems<br>Provide solutions for how two students might reduce their stress levels | Words related to managing stress<br>Forms of *confession, heal, compassionate, appreciation,* and *empathy* | TED Talk *Why Stress Is Good for You* **Kelly McGonigal**<br>Pronunciation Skill Thought groups | Analyze an infographic about different causes of stress<br>Synthesize and evaluate ideas about stress levels | **Group presentation:** Conduct a survey on stress and present the results<br>Presentation Skill Vary your pace |
| A talk about a cause<br>Use supporting arguments and emotional appeals to convince others | Words related to voluntourism<br>Antonyms of *external, inclusive, dependence,* and *outsider* | NG Explorer *A Hero on Every Coastline* **Asha De Vos**<br>Pronunciation Skill Stress inside thought groups | Analyze an infographic about the spending of charitable organizations<br>Synthesize and evaluate ideas about different ways to help | **Individual presentation:** Convince others to support an organization you believe in<br>Presentation Skill Make an emotional connection |
| A conversation about future career directions<br>Discuss and recommend a career choice | Words related to business<br>Antonyms for *tough, infinite, prior,* and *humble* | TED Talk *5 Ways to Kill Your Dreams* **Bel Pesce**<br>Pronunciation Skill Continuing and concluding | Analyze an infographic about business failure rates<br>Synthesize and evaluate ideas about the reasons for business failures | **Individual presentation:** A new business you would like to start<br>Presentation Skill Pause effectively |
| A conversation between a fitness coach and a client<br>Identify areas for improvement and suggest lifestyle changes | Words related to sleep<br>The prefix *inter-* | TED Talk *Sleep Is Your Superpower* **Matt Walker**<br>Pronunciation Skill Word endings | Analyze an infographic about factors that can contribute to a longer life<br>Synthesize and evaluate ideas about various ways to maintain health | **Group presentation:** Argue for or against a statement related to healthy living<br>Presentation Skill Organize information logically |
| A conversation about barriers faced by people with disabilities<br>Identify problems and suggest solutions | Words related to limits<br>Collocations with *encounter* | TED Talk *Embrace the Shake* **Phil Hansen**<br>Pronunciation Skill *-ed* endings | Analyze an infographic about changing mindsets<br>Synthesize and evaluate ideas about feelings and limitations | **Individual presentation:** Someone who has overcome a limitation to achieve success<br>Presentation Skill Use figurative language |

# Welcome to *21st Century Communication*, Second Edition

***21st Century Communication Listening, Speaking, and Critical Thinking*** uses big ideas from TED and National Geographic Explorers to look at one topic from different perspectives, present real and effective communication models, and prepare students to share their ideas confidently in English. Each unit develops students' listening, speaking, and critical thinking skills to achieve their academic and personal goals.

A ballet dancer waits to go on stage during a performance of *The Nutcracker* in Leeds, England.

4

Stress: Friend or Foe?

Q How do our attitudes affect us?

The dancer in this photo is about to go on stage in front of hundreds of people. For many people, this would be a stressful experience, but not everyone reacts to stress in the same way. And is stress always a bad thing? In this unit, we'll look at what stress and anxiety really are, and strategies that can help us through stressful situations.

THINK and DISCUSS

1 Look at the photo and read the caption. How do you think the dancer is feeling?

2 Look at the essential question and the unit introduction. What kind

people face today?

62

Each unit opens with an **impactful photograph** to introduce the topic and act as a springboard for classroom discussion.


Q How do our attitudes affect us?

The dancer in this photo is about to go on stage in front of hundreds of people. For many people, this would be a stressful experience, but not everyone reacts to stress in the same way. And is stress always a bad thing? In this unit, we'll look at what stress and anxiety really are, and strategies that can help us through stressful situations.


**NEW** The **Essential Question** outlines the central idea of the unit and directs students' focus to the main topic.

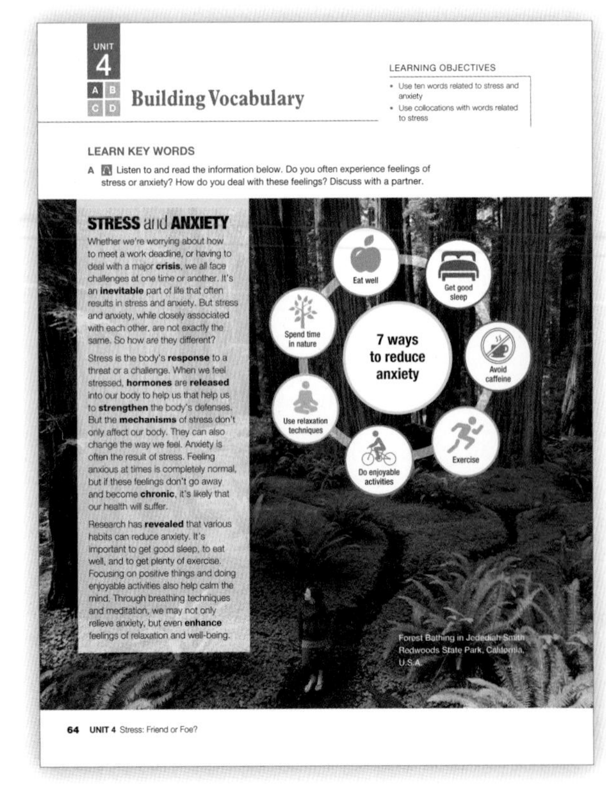

UNIT 4

A B C D Building Vocabulary

LEARNING OBJECTIVES
- Use ten words related to stress and anxiety
- Use collocations with words related to stress

LEARN KEY WORDS

A Listen to and read the information below. Do you often experience feelings of stress or anxiety? How do you deal with these feelings? Discuss with a partner.

STRESS and ANXIETY

Whether we're worrying about how to meet a work deadline, or having to deal with a major **crisis**, we all face challenges at one time or another. It's an **inevitable** part of life that often results in stress and anxiety. But stress and anxiety, while closely associated with each other, are not exactly the same. So how are they different?

Stress is the body's **response** to a threat or a challenge. When we feel stressed, **hormones** are **released** into our body to help us that help us to **strengthen** the body's defenses. But the **mechanisms** of stress don't only affect our body. They can also change the way we feel. Anxiety is often the result of stress. Feeling anxious at times is completely normal, but if these feelings don't go away and become **chronic**, it's likely that our health will suffer.

Research has **revealed** that various habits can reduce anxiety. It's important to get good sleep, to eat well, and to get plenty of exercise. Focusing on positive things and doing enjoyable activities also help calm the mind. Through breathing techniques and meditation, we may not only relieve anxiety, but even **enhance** feelings of relaxation and well-being.



Forest Bathing in Jedediah Smith Redwoods State Park, California, U.S.A.

64 UNIT 4 Stress: Friend or Foe?

**UPDATED Building Vocabulary** uses infographics and readings to introduce vocabulary in context and teach words and phrases needed for academic studies.

# Big ideas inspire many viewpoints. What's yours?

**UPDATED Viewing and Note-taking** allows students to explore one aspect of the unit theme and sharpen their academic skills with note-taking and listening comprehension practice.

**Note-taking Skill**
Using Symbols
To show relationships between ideas, it can be easier and quicker to use symbols when you take notes.

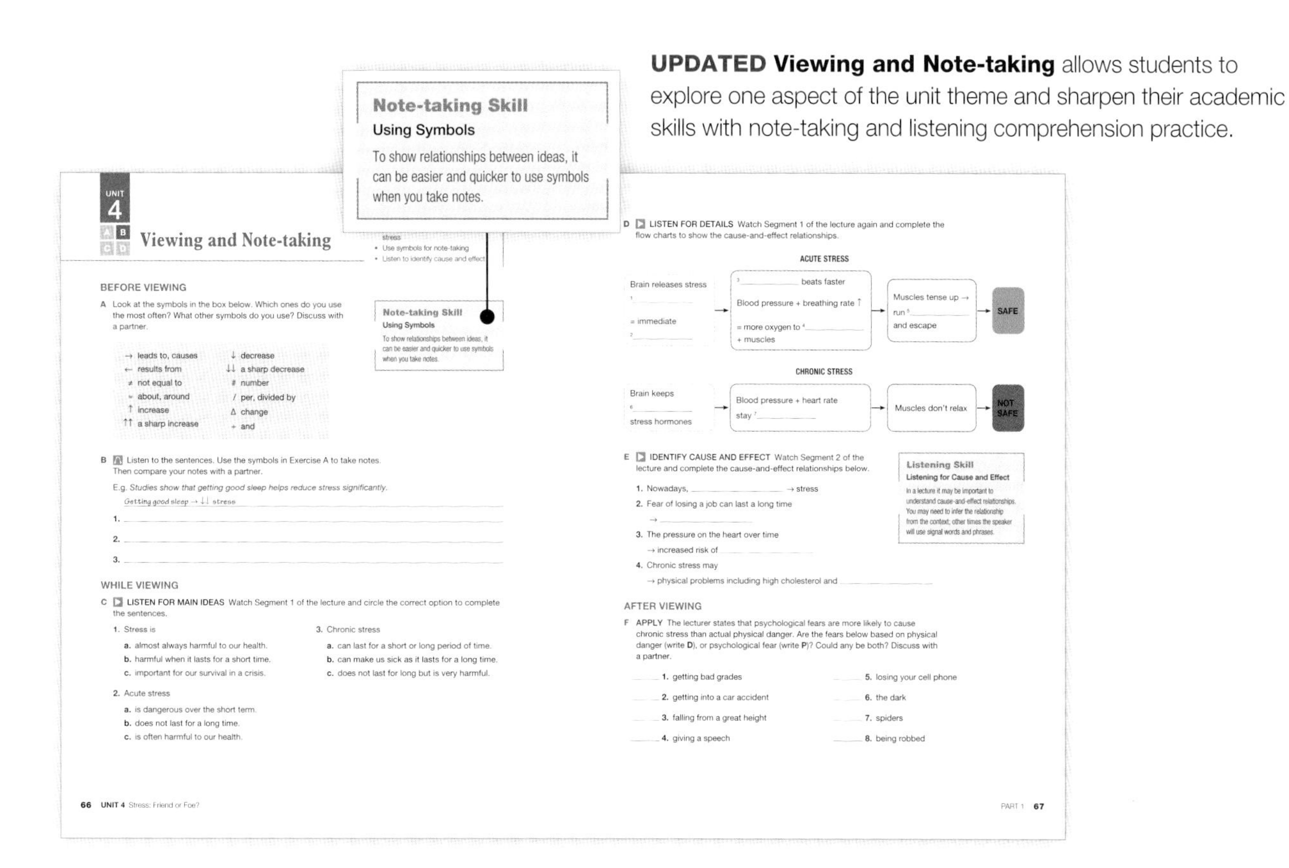

UNIT 4 A B C D

## Viewing and Note-taking

stress
- Use symbols for note-taking
- Listen to identify cause and effect

BEFORE VIEWING

A Look at the symbols in the box below. Which ones do you use the most often? What other symbols do you use? Discuss with a partner.

| | |
|---|---|
| → leads to, causes | ↓ decrease |
| ← results from | ↓↓ a sharp decrease |
| ≠ not equal to | # number |
| ≈ about, around | / per, divided by |
| ↑ increase | Δ change |
| ↑↑ a sharp increase | + and |

**Note-taking Skill**
Using Symbols
To show relationships between ideas, it can be easier and quicker to use symbols when you take notes.

B Listen to the sentences. Use the symbols in Exercise A to take notes. Then compare your notes with a partner.

E.g. *Studies show that getting good sleep helps reduce stress significantly.*
*Getting good sleep → ↓↓ stress*

1. ______
2. ______
3. ______

WHILE VIEWING

C LISTEN FOR MAIN IDEAS Watch Segment 1 of the lecture and circle the correct option to complete the sentences.

1. Stress is
   a. almost always harmful to our health.
   b. harmful when it lasts for a short time.
   c. important for our survival in a crisis.
2. Acute stress
   a. is dangerous over the short term.
   b. does not last for a long time.
   c. is often harmful to our health.
3. Chronic stress
   a. can last for a short or long period of time.
   b. can make us sick as it lasts for a long time.
   c. does not last for long but is very harmful.

D LISTEN FOR DETAILS Watch Segment 1 of the lecture again and complete the flow charts to show the cause-and-effect relationships.

ACUTE STRESS

Brain releases stress [1] ______ = immediate [2] ______ → [3] ______ beats faster / Blood pressure + breathing rate ↑ / = more oxygen to [4] ______ + muscles → Muscles tense up → run [5] ______ and escape → SAFE

CHRONIC STRESS

Brain keeps [6] ______ stress hormones → Blood pressure + heart rate stay [7] ______ → Muscles don't relax → NOT SAFE

E IDENTIFY CAUSE AND EFFECT Watch Segment 2 of the lecture and complete the cause-and-effect relationships below.

1. Nowadays, ______ → stress
2. Fear of losing a job can last a long time → ______
3. The pressure on the heart over time → increased risk of ______
4. Chronic stress may → physical problems including high cholesterol and ______

**Listening Skill**
Listening for Cause and Effect
In a lecture it may be important to understand cause-and-effect relationships. You may need to infer the relationship from the context; other times the speaker will use signal words and phrases.

AFTER VIEWING

F APPLY The lecturer states that psychological fears are more likely to cause chronic stress than actual physical danger. Are the fears below based on physical danger (write D), or psychological fear (write P)? Could any be both? Discuss with a partner.

___ 1. getting bad grades
___ 2. getting into a car accident
___ 3. falling from a great height
___ 4. giving a speech
___ 5. losing your cell phone
___ 6. the dark
___ 7. spiders
___ 8. being robbed

66 UNIT 4 Stress: Friend or Foe?

PART 1 67

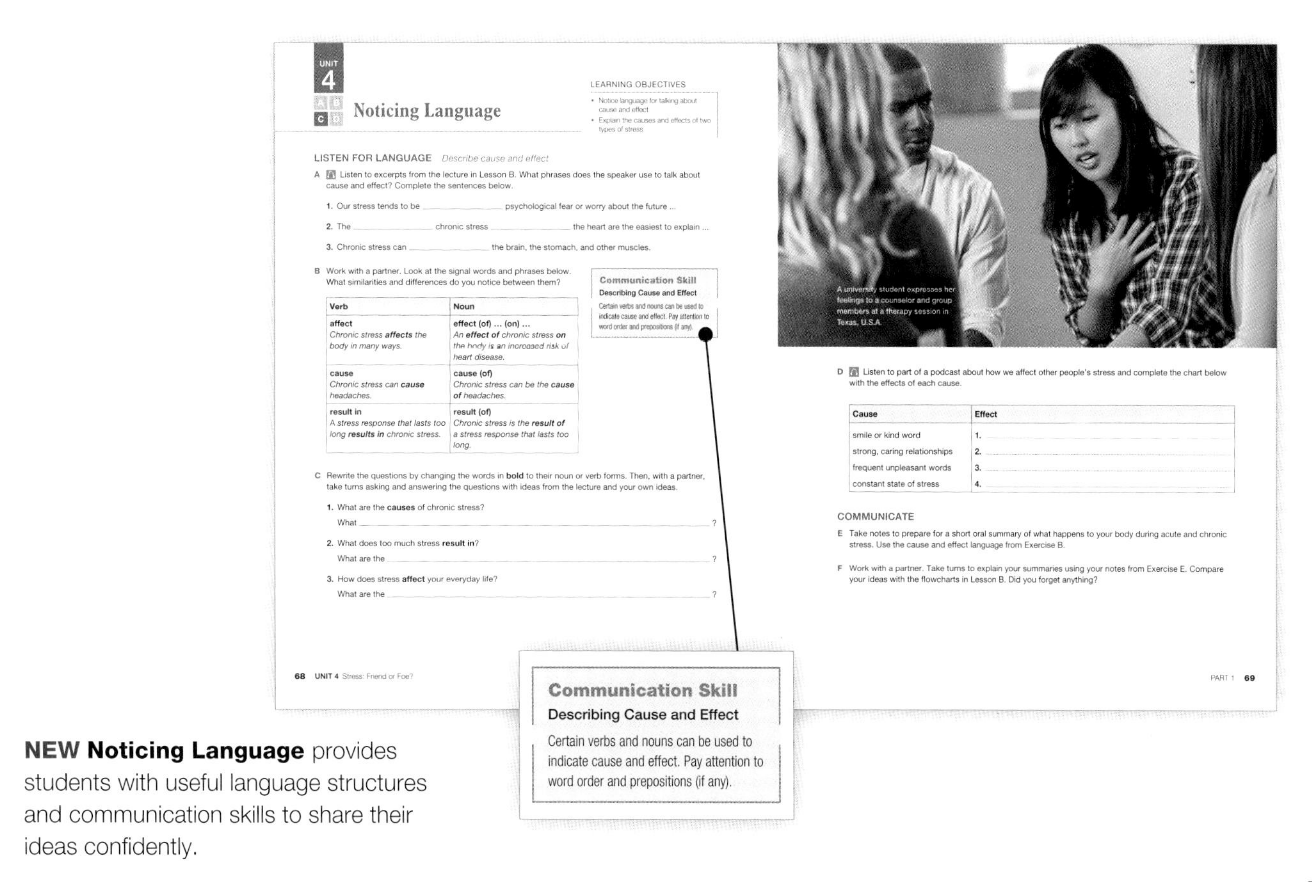

UNIT 4 A B C D

## Noticing Language

LEARNING OBJECTIVES
- Notice language for talking about cause and effect
- Explain the causes and effects of two types of stress

LISTEN FOR LANGUAGE *Describe cause and effect*

A Listen to excerpts from the lecture in Lesson B. What phrases does the speaker use to talk about cause and effect? Complete the sentences below.

1. Our stress tends to be ______ psychological fear or worry about the future ...
2. The ______ chronic stress ______ the heart are the easiest to explain ...
3. Chronic stress can ______ the brain, the stomach, and other muscles.

B Work with a partner. Look at the signal words and phrases below. What similarities and differences do you notice between them?

| Verb | Noun |
|---|---|
| **affect**<br>*Chronic stress **affects** the body in many ways.* | **effect (of) ... (on) ...**<br>*An **effect of** chronic stress **on** the body is an increased risk of heart disease.* |
| **cause**<br>*Chronic stress can **cause** headaches.* | **cause (of)**<br>*Chronic stress can be the **cause of** headaches.* |
| **result in**<br>*A stress response that lasts too long **results in** chronic stress.* | **result (of)**<br>*Chronic stress is the **result of** a stress response that lasts too long.* |

**Communication Skill**
Describing Cause and Effect
Certain verbs and nouns can be used to indicate cause and effect. Pay attention to word order and prepositions (if any).

C Rewrite the questions by changing the words in **bold** to their noun or verb forms. Then, with a partner, take turns asking and answering the questions with ideas from the lecture and your own ideas.

1. What are the **causes** of chronic stress?
   What ______?
2. What does too much stress **result in**?
   What are the ______?
3. How does stress **affect** your everyday life?
   What are the ______?

A university student expresses her feelings to a counselor and group members at a therapy session in Texas, U.S.A.

D Listen to part of a podcast about how we affect other people's stress and complete the chart below with the effects of each cause.

| Cause | Effect |
|---|---|
| smile or kind word | 1. |
| strong, caring relationships | 2. |
| frequent unpleasant words | 3. |
| constant state of stress | 4. |

COMMUNICATE

E Take notes to prepare for a short oral summary of what happens to your body during acute and chronic stress. Use the cause and effect language from Exercise B.

F Work with a partner. Take turns to explain your summaries using your notes from Exercise E. Compare your ideas with the flowcharts in Lesson B. Did you forget anything?

68 UNIT 4 Stress: Friend or Foe?

PART 1 69

**Communication Skill**
Describing Cause and Effect
Certain verbs and nouns can be used to indicate cause and effect. Pay attention to word order and prepositions (if any).

**NEW Noticing Language** provides students with useful language structures and communication skills to share their ideas confidently.

**NEW Communicating Ideas** encourages students to express their opinions, make decisions, and explore solutions to problems through collaboration.

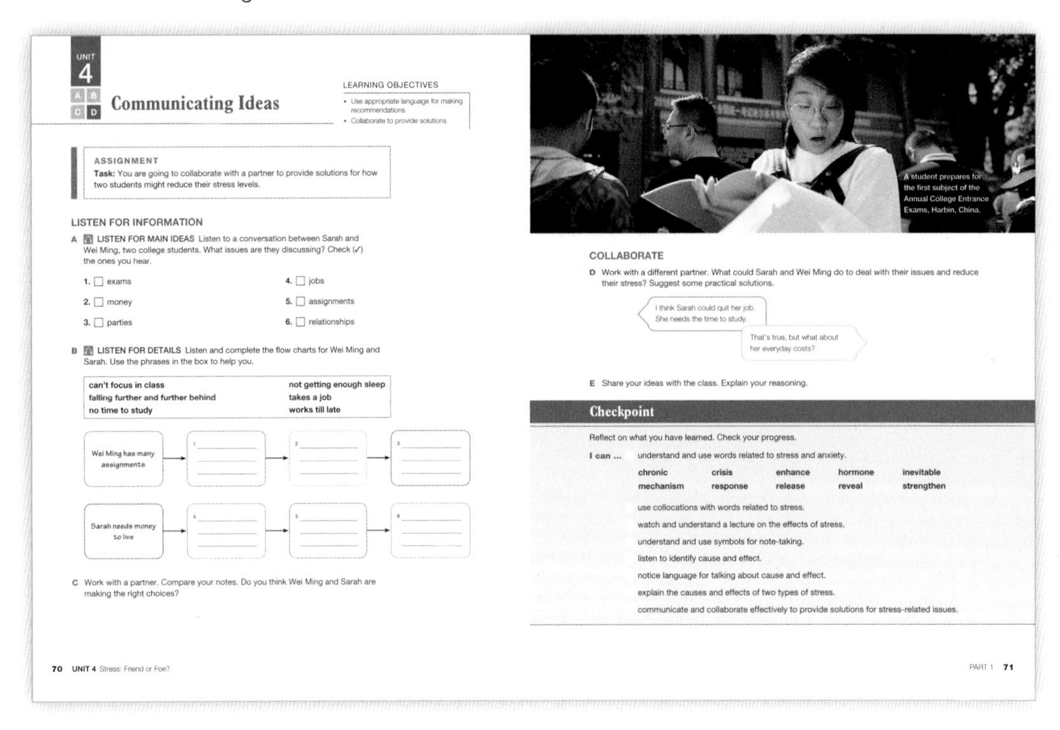

UNIT 4

A B C D

## Communicating Ideas

LEARNING OBJECTIVES

- Use appropriate language for making recommendations
- Collaborate to provide solutions

**ASSIGNMENT**

**Task:** You are going to collaborate with a partner to provide solutions for how two students might reduce their stress levels.

### LISTEN FOR INFORMATION

**A** **LISTEN FOR MAIN IDEAS** Listen to a conversation between Sarah and Wei Ming, two college students. What issues are they discussing? Check (✓) the ones you hear.

1. ☐ exams
2. ☐ money
3. ☐ parties
4. ☐ jobs
5. ☐ assignments
6. ☐ relationships

**B** **LISTEN FOR DETAILS** Listen and complete the flow charts for Wei Ming and Sarah. Use the phrases in the box to help you.

| | |
|---|---|
| **can't focus in class** | **not getting enough sleep** |
| **falling further and further behind** | **takes a job** |
| **no time to study** | **works till late** |

Wei Ming has many assignments → 1 ______ → 2 ______ → 3 ______

Sarah needs money to live → 4 ______ → 5 ______ → 6 ______

**C** Work with a partner. Compare your notes. Do you think Wei Ming and Sarah are making the right choices?

70 UNIT 4 Stress: Friend or Foe?

A student prepares for the first subject of the Annual College Entrance Exams, Harbin, China.

### COLLABORATE

**D** Work with a different partner. What could Sarah and Wei Ming do to deal with their issues and reduce their stress? Suggest some practical solutions.



**E** Share your ideas with the class. Explain your reasoning.

### Checkpoint

Reflect on what you have learned. Check your progress.

**I can ...** understand and use words related to stress and anxiety.

| | | | | |
|---|---|---|---|---|
| **chronic** | **crisis** | **enhance** | **hormone** | **inevitable** |
| **mechanism** | **response** | **release** | **reveal** | **strengthen** |

use collocations with words related to stress.
watch and understand a lecture on the effects of stress.
understand and use symbols for note-taking.
listen to identify cause and effect.
notice language for talking about cause and effect.
explain the causes and effects of two types of stress.
communicate and collaborate effectively to provide solutions for stress-related issues.

PART 1 71

UNIT 4

E F G H

## Viewing and Note-taking

LEARNING OBJECTIVES

- Watch and understand a talk about our attitudes toward stress
- Notice the use of thought groups

TEDTALKS

**TED**TALKS

Stanford University psychologist **Kelly McGonigal** is the author of a book titled *The Upside of Stress: Why Stress Is Good for You, and How to Get Good at It.* In her TED Talk, *How to Make Stress Your Friend*, she encourages us to rethink our attitudes toward stress.

### BEFORE VIEWING

**A** Read the information about Kelly McGonigal. In what ways might stress be good for us? Discuss your ideas with a partner.



PART 2 75

**UPDATED Viewing and Note-taking** uses big ideas from TED and National Geographic Explorers to present another aspect of the unit theme and help students improve their academic listening, note-taking skills, and pronunciation.

LEARNING OBJECTIVES

- Interpret an infographic about the different causes of stress
- Synthesize and evaluate ideas about stress levels

**NEW Thinking Critically** gives students a chance to synthesize, analyze, and evaluate the unit's ideas and find their voice in English.

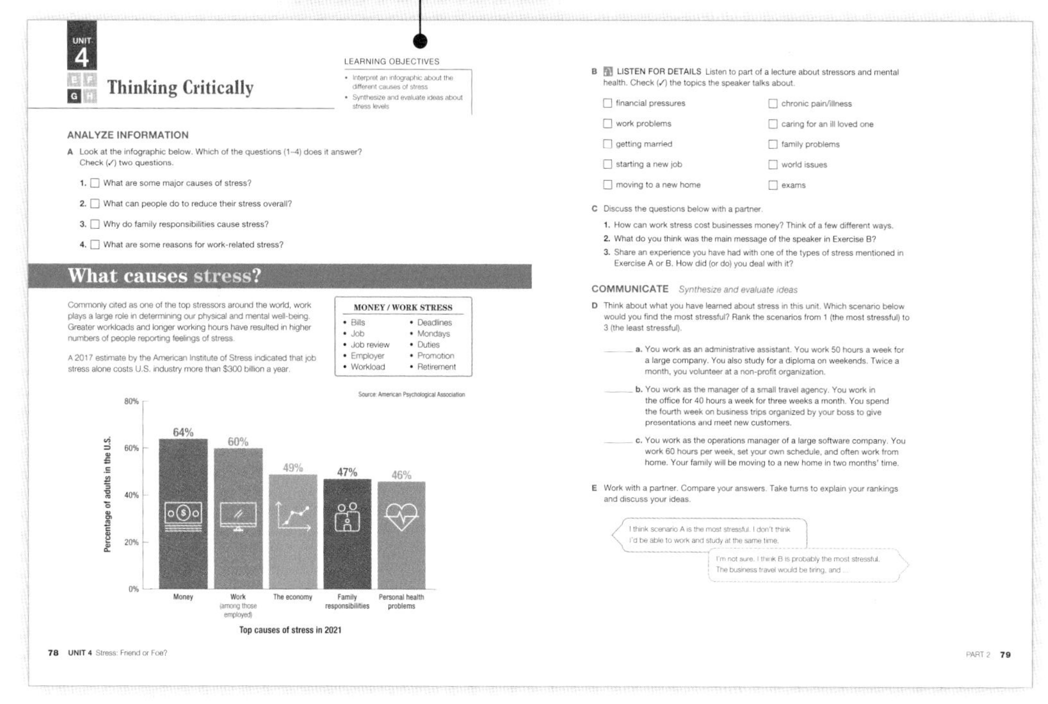

UNIT 4

## Thinking Critically

LEARNING OBJECTIVES

- Interpret an infographic about the different causes of stress
- Synthesize and evaluate ideas about stress levels

ANALYZE INFORMATION

A Look at the infographic below. Which of the questions (1–4) does it answer? Check (✓) two questions.

1. ☐ What are some major causes of stress?
2. ☐ What can people do to reduce their stress overall?
3. ☐ Why do family responsibilities cause stress?
4. ☐ What are some reasons for work-related stress?

### What causes stress?

Commonly cited as one of the top stressors around the world, work plays a large role in determining our physical and mental well-being. Greater workloads and longer working hours have resulted in higher numbers of people reporting feelings of stress.

A 2017 estimate by the American Institute of Stress indicated that job stress alone costs U.S. industry more than $300 billion a year.

MONEY / WORK STRESS

- Bills
- Job
- Job review
- Employer
- Workload
- Deadlines
- Mondays
- Duties
- Promotion
- Retirement

Source: American Psychological Association



78 UNIT 4 Stress: Friend or Foe?

B LISTEN FOR DETAILS Listen to part of a lecture about stressors and mental health. Check (✓) the topics the speaker talks about.

| | |
|---|---|
| ☐ financial pressures | ☐ chronic pain/illness |
| ☐ work problems | ☐ caring for an ill loved one |
| ☐ getting married | ☐ family problems |
| ☐ starting a new job | ☐ world issues |
| ☐ moving to a new home | ☐ exams |

C Discuss the questions below with a partner.

1. How can work stress cost businesses money? Think of a few different ways.
2. What do you think was the main message of the speaker in Exercise B?
3. Share an experience you have had with one of the types of stress mentioned in Exercise A or B. How did (or do) you deal with it?

COMMUNICATE *Synthesize and evaluate ideas*

D Think about what you have learned about stress in this unit. Which scenario below would you find the most stressful? Rank the scenarios from 1 (the most stressful) to 3 (the least stressful).

_____ a. You work as an administrative assistant. You work 50 hours a week for a large company. You also study for a diploma on weekends. Twice a month, you volunteer at a non-profit organization.

_____ b. You work as the manager of a small travel agency. You work in the office for 40 hours a week for three weeks a month. You spend the fourth week on business trips organized by your boss to give presentations and meet new customers.

_____ c. You work as the operations manager of a large software company. You work 60 hours per week, set your own schedule, and often work from home. Your family will be moving to a new home in two months' time.

E Work with a partner. Compare your answers. Take turns to explain your rankings and discuss your ideas.



PART 2 79

**UPDATED Putting It Together** has students prepare, plan, and present their ideas clearly and creatively in a final assignment.

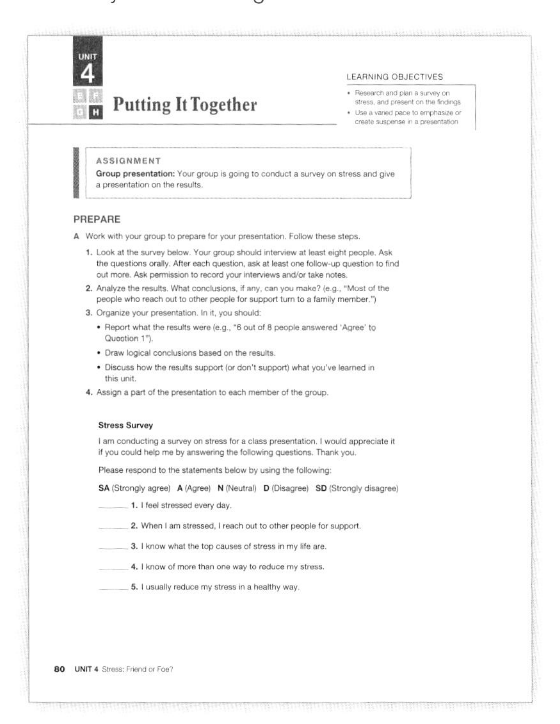

UNIT 4

## Putting It Together

LEARNING OBJECTIVES

- Research and plan a survey on stress, and present on the findings
- Use a varied pace to emphasize or create suspense in a presentation

ASSIGNMENT

**Group presentation:** Your group is going to conduct a survey on stress and give a presentation on the results.

PREPARE

A Work with your group to prepare for your presentation. Follow these steps.

1. Look at the survey below. Your group should interview at least eight people. Ask the questions orally. After each question, ask at least one follow-up question to find out more. Ask permission to record your interviews and/or take notes.
2. Analyze the results. What conclusions, if any, can you make? (e.g., "Most of the people who reach out to other people for support turn to a family member.")
3. Organize your presentation. In it, you should:
   - Report what the results were (e.g., "6 out of 8 people answered 'Agree' to Question 1").
   - Draw logical conclusions based on the results.
   - Discuss how the results support (or don't support) what you've learned in this unit.
4. Assign a part of the presentation to each member of the group.

**Stress Survey**

I am conducting a survey on stress for a class presentation. I would appreciate it if you could help me by answering the following questions. Thank you.

Please respond to the statements below by using the following:

**SA** (Strongly agree) **A** (Agree) **N** (Neutral) **D** (Disagree) **SD** (Strongly disagree)

_____ **1.** I feel stressed every day.

_____ **2.** When I am stressed, I reach out to other people for support.

_____ **3.** I know what the top causes of stress in my life are.

_____ **4.** I know of more than one way to reduce my stress.

_____ **5.** I usually reduce my stress in a healthy way.

80 UNIT 4 Stress: Friend or Foe?

The **Spark** platform delivers your digital tools for every stage of teaching and learning, including auto-graded Online Practice activities, customizable Assessment Suite tests and quizzes, Student's eBook, Classroom Presentation Tool, and downloadable Teacher's Resources.

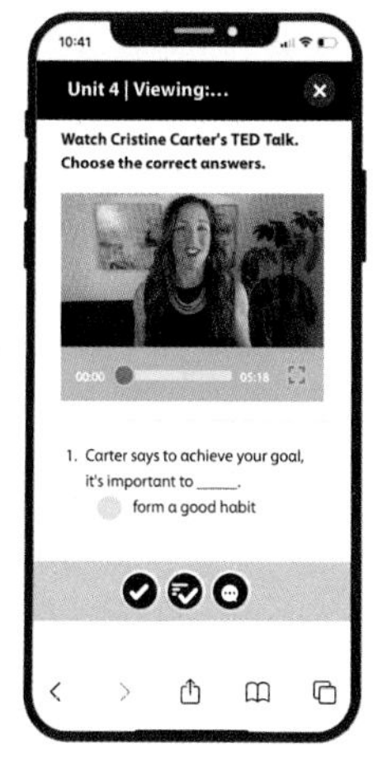



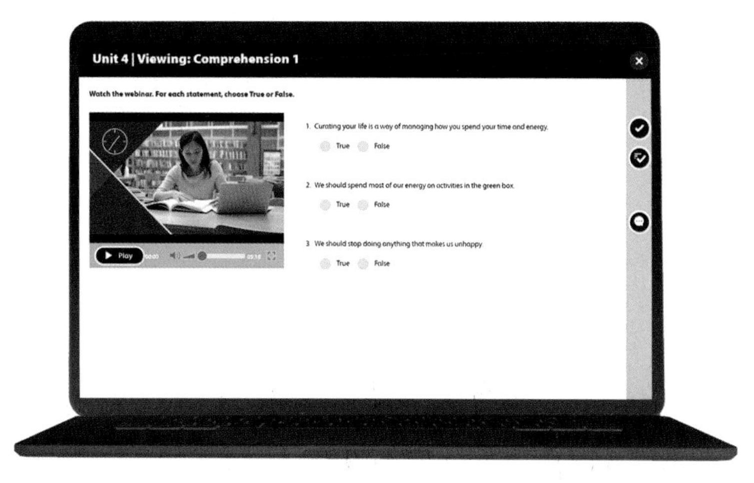

Students from Howard University celebrate their graduation. Washington, D.C., U.S.A.

# 1

# Rethinking Success

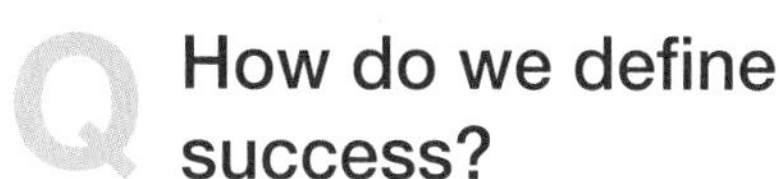

## How do we define success?

For students, success may seem easy to define. For example, the students in the photo, from Howard University in the United States, have clearly succeeded. They've graduated, and are celebrating through traditional dance. But as we mature, the nature of success changes. It becomes more complex and harder to measure. In this unit, we explore what it means to succeed, and learn how to deal with obstacles that stand in our way.

## THINK and DISCUSS

1. Look at the photo and read the caption. Would you do something like this to celebrate your success? Why, or why not?
2. Look at the essential question and the unit introduction. What do you think would make you a success in the eyes of your friends and family? Do you share their views?

UNIT 1

A B C D

# Building Vocabulary

LEARNING OBJECTIVES

- Use ten words related to success
- Use different forms of *accomplish, determination, inspirational* and *motivate*

## LEARN KEY WORDS

**A** Listen to and read the information below. What do you think about this definition of success? Do you think most people in your country would agree with it? Discuss with a partner.

## SUCCESS AND WEALTH

To many people, "**making it**" is synonymous with making a lot of money. And in many cultures, people are only considered successful when they **fulfill** society's expectations of becoming **wealthy**. As a result, the desire to become rich is what **motivates** most people to keep going. After all, money is the tool that enables us to **accomplish** all other things—from pursuing a hobby to reducing world hunger.

With money on so many people's minds, it's no wonder society celebrates individuals who go from rags to riches, and **embraces** them as **inspirational** role models. Whether they are businesspeople, celebrities, or sports figures, their stories are often similar: they start out poor, they face **failure** many times, but through hard work and **determination**, they eventually **overcome** their difficulties and succeed against the odds.

*The Money Cloud* by Ric Kasini Kadour at The Money Show, Saint Kate - The Arts Hotel, Milwaukee, U.S.A. Photo by Frank Juárez.

### Can most people get ahead if they're willing to work hard?

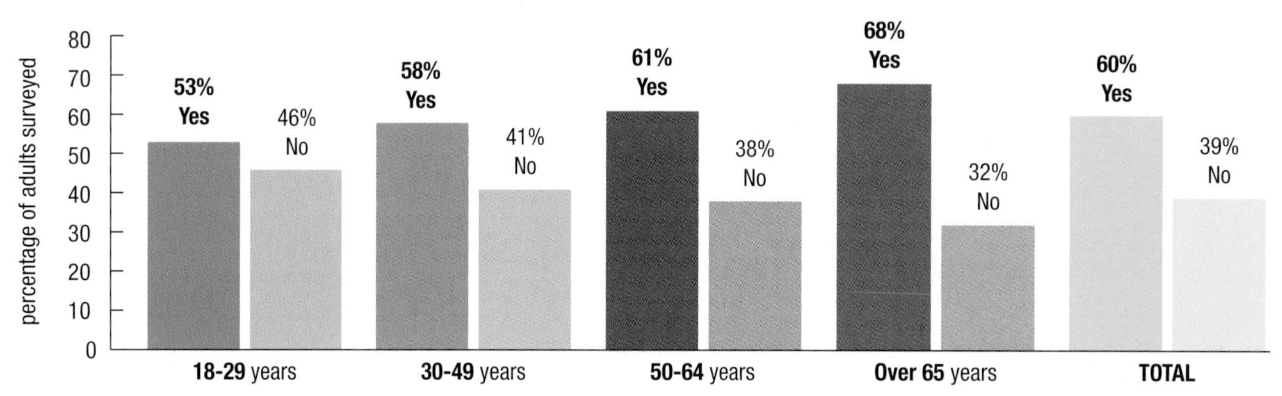


Source: Pew Research Center, survey of adults in the U.S., September 2019.

**B** Match the correct form of each word or phrase in **bold** in Exercise A with its meaning.

1. ____________________ to do something that was expected of you
2. ____________________ to achieve financial success
3. ____________________ a strong will to achieve something despite obstacles
4. ____________________ very rich
5. ____________________ to encourage or push you to do something
6. ____________________ to deal successfully with an obstacle
7. ____________________ giving hope and encouragement to other people
8. ____________________ to accept someone or something in a warm, heartfelt way
9. ____________________ to achieve something, or do something successfully
10. ____________________ a lack of success in doing or achieving something

**C** Complete the chart with the correct form of the words.

| Verb | Noun | Adjective |
|---|---|---|
| fulfill | | |
| accomplish | | |
| | | inspirational |
| | determination | |
| motivate | | |

**D** Complete the passage. using the correct form of the words in **bold** from Exercise A.

**Get Rich: 5 Tips on how to** [1]____________________

1. Work hard: be prepared to put in the hours.
2. Don't give up when faced with [2]____________________.
3. Draw [3]____________________ from those who have already become [4]____________________.
4. Learn how others managed to [5]____________________ the difficulties they faced.
5. Don't just accept the challenges you face: [6]____________________ them.

## COMMUNICATE

**E** Work with a partner. Discuss the questions below.

1. Look back at the infographic. Do the answers of people in your age group surprise you? How would you respond to the question?
2. Which idea in Exercise D do you most strongly agree with? Why?

# Viewing and Note-taking

LEARNING OBJECTIVES

- Watch a video podcast about success and failure
- Focus on main points when taking notes
- Identify main points when listening

## BEFORE VIEWING

**A** Read the passage below. Then look at the student's notes. Do you agree with the main points? Discuss with a partner.

Many people measure their success by comparing themselves to others, Unfortunately, this is usually not a good idea. Why? First, we all have different values and talents, and we face different challenges too. Comparing yourself to others is therefore like comparing apples and oranges. Second, there will always be someone who accomplishes more than you do—who runs faster, sings better, or makes more money. By always comparing yourself to those more successful than you are, you could end up feeling like a failure, no matter how much you accomplish.

> **Note-taking Skill**
> **Focusing on Main Points**
> When a speaker is explaining something, you may not have time to note down everything they say, so prioritize main ideas. If there's time in between the main ideas, note down details that support these main points. After the speaker has finished, review your notes and add in any important details you left out.

Measuring success by comparing self to others: Bad idea. Why?
1. We're all different (different talents, values, challenges)
2. feel like a failure (always someone better than you)

**B** Listen to someone talking about competition. Complete the notes.

- Learned a lesson: compete only against [1] himself.
  - When young, he was [2] great of basket ball.
    - Then, someone better came along.
    - He was the same person, but felt like a loser.
    - Lost self-confidence, quit the team.
  - In college, he [3] started basketball again.
    - His goal: not to be best player, but the best that he could be.
    - Teammates' success = his success.
    - Doing his best was good enough.
- By competing against himself, he [4] learned how to control successful ______.

## WHILE VIEWING

**C** **PREDICT** Watch Segment 1 of a video podcast about success. What do you think the speaker will talk about next?

**a.** reasons why you should believe in your ability to succeed

**b.** examples of human achievement throughout history

**c.** the importance of opportunity and luck in achieving success

**D** ▶ **CHECK PREDICTIONS** Watch Segment 2 of the video and check your prediction in Exercise C.

**E** ▶ **LISTEN FOR MAIN IDEAS** Watch Segment 2 of the video podcast again. Check (✓) two sentences that summarize the main ideas.

**1.** ☐ With enough determination, anyone can be successful.

**2.** ☑ We shouldn't divide the world into winners and losers.

**3.** ☑ Success involves both luck and opportunity.

**4.** ☐ If you have extraordinary talent, you will be successful.

> **Listening Skill**
>
> **Identifying the Main Points**
>
> It's important to identify which are a speaker's main points and which are examples. Some examples, such as case-studies or stories, can be several sentences long, and so the speaker may restate the main point after the example.

**F** ▶ **LISTEN FOR DETAILS** Watch Segment 3 of the video. Complete the notes below with main points and supporting information.

Have a balanced view of success and failure:

1. Make [1] your own definition of success.
   - Don't be afraid to [2] chang your definition own think.
2. Don't [3] measure your success against that of others.
   - (-) There'll always be [4] someone better accomplishes more thing.
3. Learn how to [5] ombrace failure.
   - We don't hear [6] what is how many live you failed before succeed.
   - (-) We get closer to success when we [7] faile.

## AFTER VIEWING

**G** **REFLECT** Read the quotes from the video. Do you agree with them, based on your personal experience or the experiences of people you know? Discuss with a partner.

**Alain De Botton**: "Never before have expectations been so high about what human beings can achieve with their life span. We're told, from many sources, that anyone can achieve anything."

**Podcast speaker:** "We all face different challenges in life, and sometimes they can't be overcome through hard work and big dreams. Success involves a lot of luck and opportunity."

**Michael Jordan:** "I've missed more than 9,000 shots in my career. I've lost almost 300 games … I've failed over and over and over again in my life. And that is why I succeed."

# Noticing Language

LEARNING OBJECTIVES

- Notice language for introducing stories
- Use someone's life story to make a point

## LISTEN FOR LANGUAGE *Use people's life stories to make a point*

**A** Listen to the following excerpts from the video podcast in Lesson B and complete the phrases the sentences below.

1. ____________ Bill Gates ____________.
2. You know ____________, ____________?
3. ____________ Michael Jordan, who many consider ...

**B** Read the tips in the gray box. Then listen to two stories from the video podcast in Lesson B and complete the notes below.

### Communication Skill

**Using People's Life Stories to Make a Point**

One way to make a point is to use real-life stories of what people have been through. When telling these stories, focus primarily on the parts that support your point. After telling the story, restate your point.

**Using people's life stories to make a point:**

- State the main point.
- Signal the beginning of the story by using expressions such as:
  *Let's take Bill Gates as an example, ...*
  *You know the story, right?*
  *Take Michael Jordan, ...*
  *You may be familiar with the story of ...*
  *Let me tell you about someone who ...*
  *Back when I was ...*
- Tell the story.
- Restate the main point.

<u>Bill Gates</u>

Main point:

- Success involves a lot of [1] ____________ and [2] ____________.

Story:

- Yes, he worked hard and was very intelligent.
- But he also went to a [3] ____________ and had unlimited access to a [4] ____________.

Restatement:

- Hard work and determination played a role in his success, but so did [5] ____________.

<u>Michael Jordan</u>

Main point:

- Learning from [6] ____________ brings us closer to [7] ____________.

Story:

- Considered by many to be the greatest basketball player of all time.
- But he also missed over [8] ____________, lost almost [9] ____________.

Restatement:

- Quote: "I've [10] ____________ over and over .... And that's why I [11] ____________."

Thomas Edison, one of the greatest inventors of all time.

**C** Listen to a person use the life story of Thomas Edison to make a point about the nature of success. Then answer the questions below.

1. What is the speaker's main point?
2. How does Edison's story help make this point?
3. How does he restate his main point at the end of the story?

**D** Work with a partner. Read the main point below. Then think of a story that illustrates the point. Practice using the story to make the point. Remember to:

- start with your main point.
- introduce your story using signal words or phrases.
- restate your main point at the end of the story.

**Main point:** Our definitions of success should change as we get older.

**Story:** ____________________________________________

**Restatement:** ____________________________________________

## COMMUNICATE

**E** Choose a main point from the box below. Think of a story to illustrate it.

- If you want to be successful, don't compete against others.
- Competing against others is the best way to become successful.
- To be successful, you need to learn from your failures.
- Success is the result of hard work and determination.
- Success involves a lot of luck.

**F** Work with a partner. Take turns making your main point with the help of a story. Follow the sequence described in Exercise B.

# Communicating Ideas

LEARNING OBJECTIVES

- Use appropriate language to discuss success
- Collaborate to create a definition of success, using a story to illustrate

**ASSIGNMENT**

**Task:** You are going to collaborate with a partner to create a definition of success, using a story example to illustrate.

## LISTEN FOR INFORMATION

**A** **LISTEN FOR MAIN IDEAS** Listen to a talk about defining success. What do you think is the main purpose of the talk?

**a.** To give examples of definitions of success

**b.** To convince people to write their own definition of success

**c.** To help people write their own definition of success

**B** **LISTEN FOR DETAILS** Listen to the talk again and complete the notes below.

How to [1] ________ in a way that's [2] ________ to you.

Think small:

- Start with [3] ________ before [4] ________.

Ask yourself meaningful questions:

1. What are your [5] ________?
2. What makes you [6] ________?
3. What [7] ________ do you need to be happy?
4. What kind of [8] ________ make you feel great?

Students in the U.K. check their examination results.

## COLLABORATE

**C** Work with a partner. Choose one of the aspects of life below. How would you define success in this aspect? Use or adapt some of the questions from the listening in Exercise B to help you in your discussion.

| friends | career | money |
|---|---|---|
| family | education | personal passions |

**D** Prepare some notes to give a short talk on your definition of success from Exercise D. Make sure you include the following:

1. Aspect of life being focused on: ____________________
2. Definition of success in this aspect: ____________________

   ____________________
3. Story to illustrate: ____________________

   ____________________
4. Restatement of main point: ____________________

> To me, success in education means developing the skills and knowledge necessary to have a fulfilling career. Take my cousin. When he was in college, he ...

**E** Work in a small group. Share your definitions of success with each other, using stories to illustrate.

# Checkpoint

Reflect on what you have learned. Check your progress.

**I can ...** understand and use words related to success.

| **accomplish** | **determination** | **embrace** | **failure** | **fulfill** |
|---|---|---|---|---|
| **inspirational** | **make it** | **motivate** | **overcome** | **wealthy** |

use different forms of *accomplish, determination, inspirational,* and *motivate.*

watch and understand a video podcast about success and failure.

focus on main points when taking notes.

identify main points when listening.

Use language for introducing stories language for introducing stories.

Use appropriate language to discuss success.

collaborate and communicate effectively to create and illustrate a definition of success

A basketball player misses two free throws in the dying seconds of a match.

# Building Vocabulary

LEARNING OBJECTIVES

- Use ten words or phrases related to opportunities and obstacles
- Understand phrasal verbs with *turn*

## LEARN KEY WORDS

**A** Listen to and read the passage below. What obstacles did Wilma Rudolph face? How did she handle them? Discuss with a partner.

**An Inspirational Athlete**

Wilma Rudolph was a successful athlete and the winner of several Olympic medals. But success didn't come easy to her. Growing up, she faced many **obstacles**.

Rudolph suffered from several illnesses, including polio, which caused her to lose strength in her left leg. She was teased and **rejected** by other children, and had to work hard for many years in order to regain her strength.

In high school, Rudolph took up basketball, but a track and field coach noticed her athleticism and was **impressed**. Under his guidance, Rudolph eventually **turned into** one of the fastest athletes in the world.

After winning three gold medals at the 1960 Olympics, Rudolph **turned down** other opportunities to compete. She became a teacher and a coach, inspiring new generations of athletes to pursue their dreams.

Wilma Rudolph shows her Olympic medals.

**B** Look at the main photo. Basketball player Anton Beard has just missed two shots that would have won the game for his team. Discuss the questions below with a partner.

1. How do you think Beard felt after missing the shots?
2. What do you think is the best way to respond to setbacks like these?
3. Have you ever failed or experienced setbacks? What was your response?

**C** Match the correct form of each word or phrase in **bold** in Exercise A with its meaning.

1. impressed having a feeling of respect for someone's achievement.
2. obstacles something that makes it difficult to do something
3. rejected to be excluded, or not allowed to be part of a group
4. turned into to change and become something else
5. turned down to say no to an opportunity or offer

**D** Read the excerpts from Jia Jiang's TED Talk in Lesson F. Choose the definition that best matches each word in **bold**.

1. "… I was presented with an **investment** opportunity, and then, I was turned down."
   - **a.** something that people buy because they expect it to increase in value
   - **b.** a job offer with better prospects that pays more money than before
   - **c.** a product or service that's offered at a discount for a limited time

2. "Would any successful **entrepreneur** quit like that? No way."
   - **a.** a person in the entertainment business
   - **b.** a person who has just gotten their first job
   - **c.** a person who owns or sets up a business

3. "And basically the idea is for 30 days you go out and look for rejection, and every day get rejected at something, and then, by the end, you **desensitize** yourself from the pain."
   - **a.** do something unpleasant so many times that it no longer bothers you
   - **b.** get rejected so many times that you lose the will to keep trying
   - **c.** repeat something so many times that you begin to lose interest in it

4. "The next day, no matter what happens, I'm not going to run. I'll stay **engaged."**
   - **a.** involved  **b.** scared  **c.** alone

5. "I turned around, and I just ran. I felt so **embarrassed**."
   - **a.** interested  **b.** ashamed  **c.** pressured

**E** The verb *turn* can be used to form many phrasal verbs. Complete the sentences with the phrasal verbs in the box. Use a dictionary to check your answers.

| **turn around** | **turn down** | **turn into** | **turn out** |
|---|---|---|---|

1. Successful people can ______________ a bad situation ______________ an opportunity.
2. I worry that the bank is going to ______________ me ______________ for that loan.
3. Don't worry. Everything will ______________ fine in the end.
4. I know the company's finances are not in good shape, but I'm confident we can ______________ the business ______________.

## COMMUNICATE

**F** Work with a partner. Take turns discussing the items below. Respond to your partner's ideas or ask follow-up questions.

Steve Jobs, the founder of Apple, is an entrepreneur I admire.

Why do you admire him?

1. an **entrepreneur** that you admire
2. a time when you **turned down** an opportunity

UNIT 1

E F G H

# Viewing and Note-taking

LEARNING OBJECTIVES

- Watch and understand a talk about overcoming rejection
- Notice the use of intonation in lists



## TEDTALKS

**Jia Jiang** is an entrepreneur, best-selling author, and owner of the website *Rejection Therapy*. In his TED Talk, *What I Learned from 100 Days of Rejection*, he shares his lifelong fear of rejection and his unique way of overcoming it.

### BEFORE VIEWING

**A** Read the information above about Jia Jiang. How do you think 100 days of rejection would feel? How could it benefit someone? Discuss with a partner.

## WHILE VIEWING

**B** ▶ **LISTEN FOR MAIN IDEAS** Watch Segment 1 of Jia Jiang's TED Talk and take notes. Check your predictions in Exercise A, and answer the questions below.

1. What obstacle did he face as an entrepreneur?

   ______________________________

   ______________________________

2. What did he realize his problem was?

   ______________________________

   ______________________________

3. What did he decide to do about it?

   ______________________________

   ______________________________

**C** ▶ **LISTEN FOR MAIN IDEAS** Watch Segment 2 of the TED Talk. Complete the chart below with notes about the four things Jiang did to face rejection.

| **Example 1**<br>Goal:<br>Result: | **Example 2**<br>Goal:<br>Result: |
|---|---|
| **Example 3**<br>Goal:<br>Result: | **Example 4**<br>Goal:<br>Result: |

**D** **LISTEN FOR DETAILS** What lesson did Jiang learn from each experience? Match the examples 1–4 from Exercise D with the lessons below.

_____ Being persistent about what we want can eventually lead to success.

_____ Staying engaged can help us to push through the fear of rejection.

_____ Asking "why" after we have been rejected could help us to succeed eventually.

_____ It's important not to run away when we face rejection.

---

**WORDS IN THE TALK**

*rah-rah* (adj) enthusiastic

*The Sixth Sense* (title) a movie about a boy who talks to ghosts

## AFTER VIEWING

**E** **REFLECT** Work with a partner. Discuss the questions.

1. Have you ever faced rejection before? Did you respond more like Jiang before rejection therapy or after?
2. Do you think rejection therapy works for everyone? Are there any possible negative effects?
3. What are some other feelings that get in the way of success that you think we could desensitize ourselves to?

## PRONUNCIATION *Intonation in Lists*

> **Pronunciation Skill**
> **Intonation in Lists**
> When speakers list things, they usually use a rising intonation for each item except the last one, which usually has a falling intonation. Sometimes, they use a rising intonation for the last item, to suggest that the list of items goes on.

**F** Listen to and read two excerpts from Jia Jiang's TED Talk. Underline the items he lists in each excerpt. Does he use a rising or falling intonation for the final item in each list? Why?

1. ... and that was the longest walk of my life— hair on the back of my neck standing up, I was sweating, and my heart was pounding—and I got there and said ...
2. In fact, he invited me to explain myself. And I could've said many things. I could've explained, I could've negotiated. I didn't do any of that. All I did was run.
3. But had I left after the initial rejection, I would've thought, well, it's because the guy didn't trust me, it's because I was crazy, because I didn't dress up well, I didn't look good ... It was none of those.

**G** Work with a partner. Read the sentences and underline the listed items. Draw ↗ for the items with rising intonation and ↘ for the items with falling intonation. Take turns reading the sentences out loud. Correct each other's intonation as necessary.

1. I told my friends, my family, even my colleagues. I got mixed feedback.
2. Rejection makes me feel sad, embarrassed, angry ... all sorts of bad emotions.
3. I never fear things like rejection, embarrassment, failure ... I fear not trying.

**H** Complete the sentences with ideas of your own. Draw arrows to mark rising and falling intonation. Then work with a partner and take turns reading the sentences out loud. Correct each other's intonation as necessary.

1. ____________, ____________, and ____________ are common obstacles to success.
2. I lose my self-confidence when I experience ____________, ____________ or ____________.
3. To overcome such obstacles, I will ____________, ____________, and ____________.

UNIT 1

E F G H

# Thinking Critically

LEARNING OBJECTIVES

- Interpret an infographic comparing the goals of two different generations
- Synthesize and evaluate ideas about success and failure

## ANALYZE INFORMATION

**A** Look at the infographic about millennials and Generation Z. What does it show? Check (✓) the correct options.

1. ☐ What they want to achieve before they're 30
2. ☐ Their goals, ranked from most to least important
3. ☐ Their goals, ranked from most to least common
4. ☐ How their goals change as they grow older

**B** Does any of the information in the infographic stand out as interesting or surprising? Why? Discuss with a partner.

## What do you want to accomplish before the age of 30?

Perhaps unsurprisingly, people from different generations have different ideas about what they would like to accomplish by the time they are 30. Why are their answers different? And what does this information tell us about their definitions of success?

**MILLENNIALS (people born 1981-1996)**

|  |  |
|---|---|
| Become financially independent | 59% |
| Finish my education | 52% |
| Start a career | 51% |
| Find out who I really am | 40% |
| Follow my dreams | 31% |
| Get married | 28% |
| Enjoy life while I'm young | 24% |

**GEN Z (people born 1997-2012)**

|  |  |
|---|---|
| Finish my education | 66% |
| Start a career | 66% |
| Become financially independent | 65% |
| Follow my dreams | 55% |
| Enjoy life while I'm young | 38% |
| Find out who I really am | 31% |
| Get married | 20% |

Survey population:
1,400 U.S. teens aged 13–18 in 2016
1,000 U.S. adults aged 18–29 in 2013
Source: barna.com, 2018

**C** Work in pairs. Work out the percentage differences in the popularity of the goals across the two generations.

| Goals | % diff |
|---|---|
| 1. Finish education | ______ |
| 2. Start a career | ______ |
| 3. Financial independence | ______ |
| 4. Follow my dreams | ______ |
| 5. Enjoy life while young | ______ |
| 6. Find out who I really am | ______ |
| 7. Get married | ______ |

**D** Look at the information in Exercise C. Which goal has the biggest percentage difference? Which has the smallest? What do you think this says about the two generations?

**E** Do you belong to either of the two generations in the infographic? How would you rank the goals in order of importance? Are any of your own goals missing from the infographic? Discuss with a partner.

## COMMUNICATE *Synthesize and evaluate ideas*

**F** The ideas below were all covered earlier in the unit. Check (✓) the ideas that you think were reflected in the infographic as well. Discuss your answers with a partner.

1. ☐ Money is the most important measure of my success.
2. ☐ There's much more to success than wealth and money.
3. ☐ We shouldn't compare our success to that of others.
4. ☐ Luck plays an important role in everyone's success.
5. ☐ We should all have our own definitions of success.
6. ☐ If we work hard to overcome obstacles, we will succeed.
7. ☐ Our fear of rejection is a bigger problem than actual rejection.
8. ☐ As our priorities change, so should our definitions of success.

**G** Work with a partner. Look at the eight ideas about success in Exercise F again. Which three do you most strongly agree with and why?

> I think that most people measure success by how much money they make. I do too, because money makes all my other dreams possible.

> I guess that's true for many people, but not me. For me, success means being a good person who helps improve the lives of others.

UNIT 1

E F G H

# Putting It Together

LEARNING OBJECTIVES

- Research, plan, and present on someone's success
- Use humor to connect with an audience

**ASSIGNMENT**

**Individual presentation:** You are going to give a presentation about two people you consider successful. What makes them successful, and what can we learn from them?

## PREPARE

**A** Review the unit. Work with a partner. Who are some of the well-known successful people covered in the unit? What makes each person's success story interesting or noteworthy?

**B** Think of five or six people who have been successful in different aspects of life. You can note down the names of people you know personally or search online for examples of successful people.

______________________________

______________________________

**C** Plan your presentation. Choose two people from your notes in Exercise B who defined success differently, and who achieved success in different ways. Use the chart below to help you.

| | **Person 1:** ______ | **Person 2:** ______ |
|---|---|---|
| **Aspect of life** | | |
| **Obstacles** | | |
| **How long to achieve success** | | |
| **Key to success** | | |
| **Lessons we can learn from them** | | |

**D** Look back at the vocabulary, pronunciation, and communication skills you've learned in this unit. What can you use in your presentation? Note any useful language below.

**E** Below are some ways that Jia Jiang uses humor to connect with his audience. What makes the jokes funny? Then look back at your notes in Exercise C. Think about where you could add humor to your own presentation. Note it down in your plan.

*"Then I came up with a bunch of "rah-rah" inspirational articles about "Don't take it personally, just overcome it." Who doesn't know that?"*

*"This is me—you probably can't see, this is a bad picture. You know, sometimes you get rejected by lighting, you know?"*

**Presentation Skill**

**Using Humor**

Humor can help you connect with the audience. One way to do this is by gently making fun of yourself, as Jia Jiang does in his TED Talk. You could also include a joke, use a funny image or quote, or make a comical comparison.

**F** Practice your presentation. Make use of the presentation skill that you've learned.

## PRESENT

**G** Give your presentation to a partner. Watch their presentation and evaluate them using the Presentation Scoring Rubrics at the back of the book.

**H** Discuss your evaluation with your partner. Give feedback on two things they did well and two areas for improvement.

# Checkpoint

Reflect on what you have learned. Check your progress.

**I can ...** understand and use words related to opportunities and obstacles.

| | | | | |
|---|---|---|---|---|
| **desensitize** | **embarrassed** | **engaged** | **entrepreneur** | **impressed** |
| **investment** | **obstacle** | **rejected** | **turn down** | **turn into** |

- use phrasal verbs beginning with *turn.*
- watch and understand a talk about overcoming rejection.
- use appropriate intonation when listing things.
- interpret an infographic comparing the goals of two different generations.
- synthesize and evaluate ideas about success and failure.
- use humor to connect with an audience.
- give a presentation about two people's success stories.

Tourists help repair a damaged coral reef near the island of Mo'orea, French Polynesia.

# 2

# Changemakers

## Q How can we make a difference in the world?

Most people want to do good—the question is often how. We all lead busy lives, and it's easy to feel like time isn't on our side. But there are always ways to contribute. For example, the two tourists in the photo are giving their time to help the conservation group Coral Gardeners repair damaged reefs in French Polynesia. In this unit, we look at some of the many ways people have made a difference, and also explore practical ways we can all give something back.

## THINK and DISCUSS

1 Look at the photo and read the caption. What kind of change do you think the people are trying to make?

2 Look at the essential question and the unit introduction. Who are some people who have made a positive difference in the world? What did they do?

UNIT 2

A B C D

# Building Vocabulary

LEARNING OBJECTIVES

- Use ten words related to making a difference
- Use the suffix *-tion*

## LEARN KEY WORDS

**A** Listen to and read the information. Discuss the questions below with a partner.

1. How is Encore different from most other charities?
2. How did younger doctors benefit from the project described?
3. Look at the pie chart. Which two age groups in the U.S. volunteer the most? Does this surprise you?

## PASSING THE TORCH

Changemaking is often **associated with** young people on a **mission** to change the world. But young people are not the only ones who can make an **impact**. Society has much to gain from those with more experience.

One organization that **aspires** to make the most of experience is Encore, an **innovative** charity that gets people from different generations to work together in meaningful ways. For example, for one of its projects, it paired retired doctors up with younger ones and sent them out to treat patients **on the front lines**, in places with limited healthcare access. The senior doctors shared their knowledge and expertise with their younger counterparts, who in turn gained the skill and **confidence** they needed to better perform their duties.

What is the **intention** of programs like this one? The guiding **principle** is simple: to **ensure** that the knowledge accumulated over the years by those with experience isn't wasted; that it is instead passed on to the next generation of changemakers.

**Age of Volunteers in the U.S.**

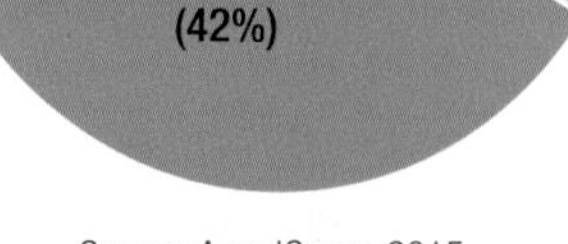


Source: AmeriCorps, 2015

**B** Match the correct form of each word or phrase in **bold** in Exercise A with its meaning.

1. aspires ________ to aim to achieve something great
2. associated with ________ connected to
3. intention ________ an aim or purpose
4. confidence ________ belief in your own abilities
5. impact ________ a powerful effect
6. innovative ________ creative and original
7. mission ________ an important assignment
8. ensure ________ to make sure something happens
9. principle ________ an idea that forms the basis of something
10. on the front lines ________ where important and difficult things are happening

**C** The suffix *-tion* is often used to change verbs into nouns but there may be other spelling changes as well. Complete the sentences with the noun forms of the words in the box. Use a dictionary if necessary.

| **innovate** | **aspire** | **intend** |
|---|---|---|

1. The organization has publicly announced its intantion ________ to move its headquarters to Singapore.
2. Her clever innovation ________ have helped make farming more efficient for the locals.
3. As a child, his aspiration ________ was to become a doctor, but later he decided to focus on business instead.

**D** Read the profile of a changemaker below. Complete the passage with the correct form of the words in **bold** from Exercise A.

Ralph Nader's name is closely [1] associate ________ with consumer rights. Early in his career, he became famous for fighting against American auto companies making unsafe products. Throughout his career, Nader's [2] intention ________ remained the same: to protect the "little guy" from corporate America. While he wasn't the first to fight for this cause, his methods were [3] innovative ________, and his [4] aspiration ________ can still be felt today.

## COMMUNICATE

**E** Work with a partner. Discuss the questions below.

1. What are some of the most important issues faced by society today?
2. If you became a changemaker, what would your mission be?

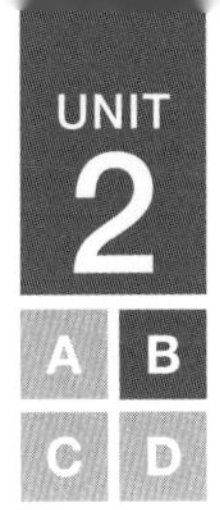

# Viewing and Note-taking

LEARNING OBJECTIVES

- Watch and understand a lecture on making a difference
- Use a concept map when taking notes
- Ask questions while listening

## BEFORE VIEWING

**A** You are going to watch a lecture about our career choices and making an impact. Which of these jobs do you think allow you to make the biggest impact? Rank them from 1 (most impact) to 6 (least impact). Discuss with a partner.

_______ marine scientist

_______ investment banker

_______ doctor

_______ environmental activist

_______ lecturer

_______ lawyer

**B** Listen to four excerpts from the lecture. At the end of Excerpts 1, 2, and 3, write a question based on what you think the speaker will say next. Were your questions answered?

**1.** ______________________________

**2.** ______________________________

**3.** ______________________________

### Listening Skill

**Asking Questions While Listening**

As you listen to a talk or lecture, try to predict what the speaker will talk about next. One way to listen ahead is by asking yourself *Wh-* questions (*Who, What, When, Where, How,* and *Why*). You won't always get the answers, but keeping your questions in mind can help you better evaluate a speaker's message.

## WHILE VIEWING

**C** **LISTEN FOR MAIN IDEAS** Watch Segment 1 of the lecture. Which of the statements below best describes the main idea of the talk?

**a.** Imogen Napper's path is one that all aspiring changemakers should follow.

**b.** One way of being a changemaker is through the career you choose.

**c.** Being a changemaker means seeing a problem in society and working to solve it.

**D** Complete the the concept map on the next page with details you remember from Segment 1 of the lecture. Compare your concept map with a partner. Did you miss anything?

**E** **LISTEN FOR DETAILS** Watch Segment 1 again and check your answers.

### Note-taking Skill

**Using a Concept Map**

A concept map is similar to a mind map, but its main purpose is to show the connections between ideas. Arrows are often used to show how ideas are connected to one another, for example, by linking main ideas with their supporting details.

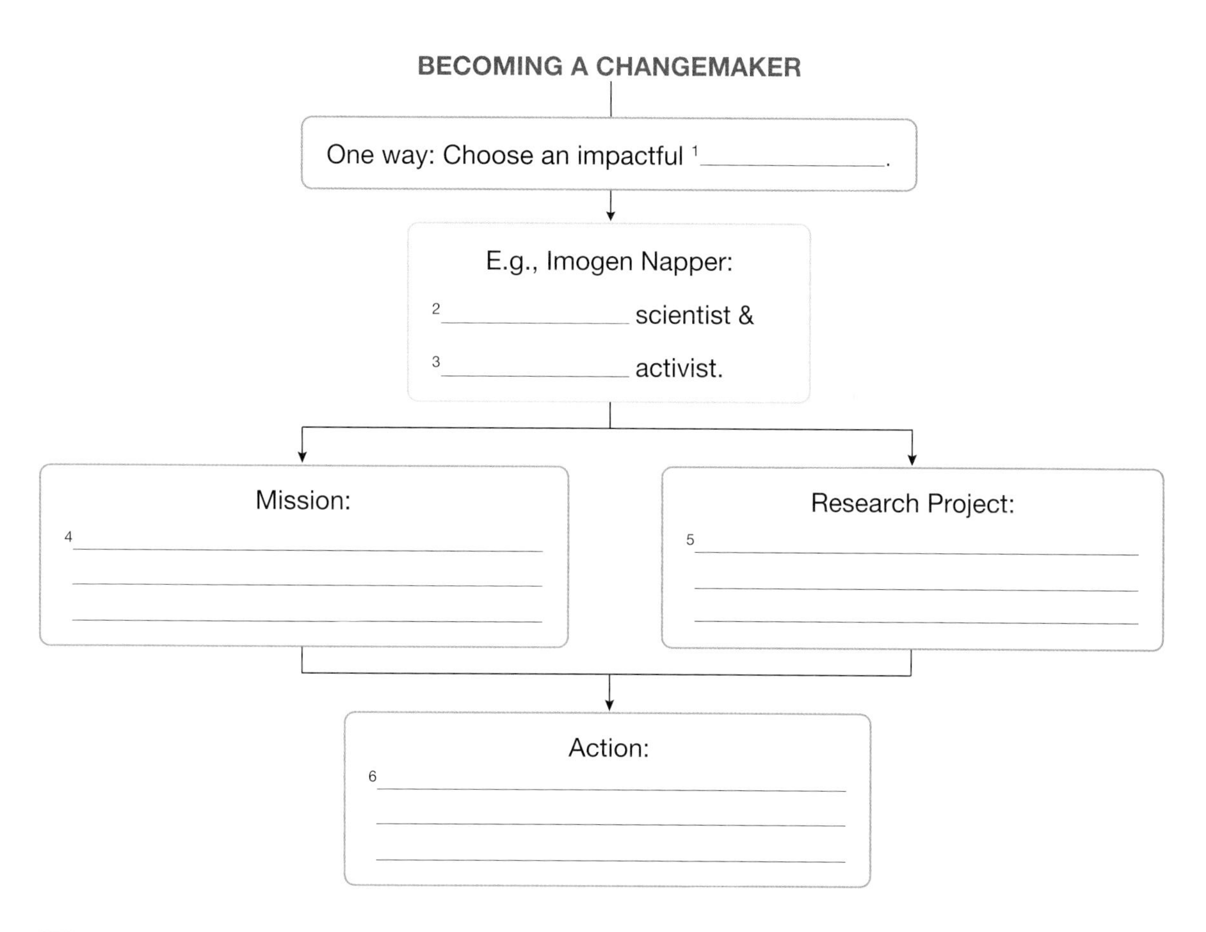


**F** **LISTEN FOR DETAILS** Watch Segment 2 of the lecture and complete the notes below. Do you do you think this is a fair comparison? Discuss with a partner.

| Investment Banker | Doctor |
|---|---|
| Method of helping: | Method of helping: |
| Lives saved: | Lives saved: |
| Source/evidence: | Source/evidence: |

**G** **INFER** Watch Segment 3 of the lecture. Check (✓) the points the lecturer would probably agree with.

**a.** ☐ On average, doctors save about 200 lives per year.
**b.** ☐ You don't have to be a front-line worker to save lives.
**c.** ☐ If we really want to make a difference, our career choices matter.
**d.** ☐ It's OK to choose a career just for the money.

## AFTER VIEWING

**H** **REFLECT** Discuss the questions below with a partner.

1. Do you think career choice matters when it comes to making a difference? Why, or why not?
2. Do you think sending a check is as meaningful as doing front-line charity work? Why, or why not?

# Noticing Language

LEARNING OBJECTIVES

- Notice language for seeking and offering clarification
- Discuss opinions about making a difference

## LISTEN FOR LANGUAGE *Seek and offer clarification*

**A** Listen to the following excerpts from the lecture in Lesson B. Complete the phrases used to seek and offer clarification.

**1.** Ralf: I'm sorry, ____________. Shouldn't the average number of lives doctors save be 5 times 40 … ?

**2.** Nadine: I'm not sure about Singer's idea. ____________ front-line work is not important … ?

**3.** Lara: No, ____________. He's saying that we can't all be volunteers …

**4.** Nadine: I see. ____________, he's not saying that the work that doctors and volunteers do *isn't* important.

### Communication Skill

**Seeking and Offering Clarification**

Listeners often need to clarify what they hear, and speakers need to respond appropriately. There are expressions you can use to ask for clarification, restate main points, or directly state implied points.

**B** Read the phrases in the box. Discuss with a partner.

**1.** Which of the phrases are for seeking clarification (write **S**), and which are for offering clarification (write **O**)?

**2.** Which of the phrases for seeking clarification is a request for more specific and detailed information?

**3.** Which of the phrases for seeking clarification do you use just before paraphrasing what a speaker has said?

**4.** Which of the phrases for offering clarification are responses to statements that are only partly true?

___ I think I'm missing something.
___ No, not quite.
___ What he's really saying is …
___ In a sense.
___ Exactly.
___ Is he saying that … ?
___ I don't quite follow.
___ So if I understand you correctly, …
___ Could you elaborate on that a little?
___ Not at all.

**C** Listen to a conversation between Johan and Viv about bullying. Answer the questions.

**1.** What does Johan say is the only solution to bullying?

**2.** What does Viv say is the most important thing we can do to help deal with the problem of bullying?

**3.** What has Morgan Guess's foundation done to help deal with the problem of bullying?

A group of volunteers cleans up a street in Tokyo, Japan.

**D** Read the opinions below. Respond including phrases for seeking clarification.

1. I know people who volunteer and do *actual* good in the world, but I know many more who just talk about doing good things. These armchair activists are really loud on social media, but they don't actually do anything.

   ______________________________

   ______________________________

2. You know how many celebrities support environmental causes? They say that they care about climate change, yet they fly hundreds of thousands of miles every year, often on private jets. I just don't buy it.

   ______________________________

   ______________________________

**E** Work with a partner. Listen to their requests for clarification from Exercise D, and respond. Include phrases for offering clarification.

### COMMUNICATE

**F** Choose one of the opinions below. Do you agree? Why, or why not? Take notes.

> **1.** Armchair activism doesn't make a difference in the real world.
> **2.** Most celebrities support causes because it improves their image.
> **3.** It's wrong to work in an industry that does harm to the world.

**G** Find a partner who chose a different opinion in Exercise F. Discuss your positions and reasons. Use phrases for seeking and offering clarification.

Armchair activism certainly isn't front-line work, but it does raise awareness …

Supporting a cause may improve a celebrity's image, but that doesn't mean …

# Communicating Ideas

## LEARNING OBJECTIVES

- Use appropriate language for seeking and offering clarification
- Collaborate to create a proposal to help the local community

**ASSIGNMENT**

**Task:** You are going to collaborate with a partner to share ideas about how to make a difference in your community.

## LISTEN FOR INFORMATION

**A** **LISTEN FOR MAIN IDEAS** Listen to a conversation between Aisha and Wayne. Discuss the questions below with a partner.

**1.** Are Aisha and Wayne students, professors, or business people? Why, do you think?

________________________________________

________________________________________

**2.** What does Aisha want to do to help her local community?

________________________________________

________________________________________

**B** **LISTEN FOR DETAILS** Work with a partner. Try to complete the notes below. Then listen to check your ideas.

*Local Council New Initiatives Scheme*

*The council listens to new ideas that can help the [1] ____________ – proposals for new [2] ____________.*

***What is Aisha's idea?*** *To help the community through [3] ____________.*

***Who exactly will it help?*** *Unemployed young people from [4] ____________ backgrounds.*

***How exactly will it help them?*** *Giving them food service training and then [5] ____________.*

***What other benefits are there?*** *It can help with staffing and may be good [6] ____________.*

***Is there potential for growth?*** *Could expand into other sorts of work skills training, e.g., office skills, basic [7] ____________, etc.*

***What are the first steps?*** *Start small—Aisha's parents can provide some [8] ____________ and work experience at their café. Later, other local cafés and restaurants can get involved.*

**C** **Work with a partner.** If you were on the local council, would you support Aisha's project? What questions would you ask before making a decision?

## COLLABORATE

**D** Work with a partner. How would you like to help your local community? Choose one of the ideas from the box in the centre of the mind map or add your own. Then discuss and take notes in the other boxes.

**1. Who exactly will it help?**

**2. How exactly will it help them?**

- ☐ cleaning up the streets
- ☐ feeding the hungry
- ☐ preventing school bullying
- ☐ helping the homeless
- ☐ your idea: ______________________

**3. What other benefits are there?**

**4. What are the first steps?**

**E** Work with another pair. Take it in turns to present your ideas as if you were speaking to your local council. Use language to seek and offer clarification as needed.

# Checkpoint

Reflect on what you have learned. Check your progress.

**I can ...** understand and use words and phrases related to making a difference.

| | | | | |
|---|---|---|---|---|
| **aspire** | **associated with** | **confidence** | **ensure** | **impact** |
| **innovative** | **intention** | **mission** | **on the front lines** | **principle** |

use the suffix *-tion*.

watch and understand a lecture on making a difference.

use a concept map when taking notes.

ask questions while listening.

notice language for seeking and offering clarification.

use clarifying language to discuss opinions about making a difference.

collaborate and communicate effectively to create a proposal to help the local community.

A quiet moment of sharing between two capuchin monkeys in Costa Rica.

UNIT 2

E F G H

# Building Vocabulary

LEARNING OBJECTIVES

- Use ten words related to equality
- Use the suffix *-or*

## LEARN KEY WORDS

**A** Listen to and read the passage below. What was exciting about Brosnan and de Waal's experiment? Discuss with a partner.

**Who cares about fairness?**

What makes people want to create change in the world around them? One of the biggest motivations is often to take action against injustice.

Living in any society requires a degree of **collaboration**. Laws and unspoken rules help people to **navigate** interactions with other people. But when people feel that they are being treated unfairly, they may no longer wish to follow the rules.

Humans have a strong sense of fairness. For us, it is **essential** to recognize not only when we are being treated worse than others, but also when we have an unfair advantage over others. This is why even young children learn quickly that sharing is important and that games should give everyone an equal chance.

Until recently, this sense of fairness was believed to be unique to humans. Then, in 2003, an experiment by scientists Sarah Brosnan and Frans de Waal provided evidence that these behaviors may **extend** to some of our animal relatives.

Brosnan and de Waal observed that capuchin monkeys paid close attention not just to the rewards that they received for completing certain tasks, but also to the rewards that other monkeys received. One monkey famously threw away her cucumber when she saw that her neighbor was receiving grapes, their preferred food.

To the researchers, these behaviors **conveyed** the monkeys' dislike of unfair treatment, but others called for stronger evidence that the monkeys would give up an advantage in the the interest of fairness. The research continues, but what we know for sure is that some humans will fight against inequality even at great cost to themselves.

**B** Work with a partner. Discuss the questions below.

1. Have you seen animals cooperating like the monkeys in the photo? Do you think it's possible that animals act out of fairness?
2. If you had an advantage over someone else, would you give it up so everyone could be equal?

**C** Match the correct form of each word in **bold** from Exercise A with its meaning.

1. ______________ working together to produce something
2. ______________ to find a way through a complex situation
3. ______________ to express a thought, feeling, or idea
4. ______________ to stretch or add to something
5. ______________ extremely important or necessary

**D** Read the excerpts from Melati Wijsen's TED Talk in Lesson F. Choose the options that are closest to the meanings of the words in **bold**.

1. In the last few years, I have spent more time in other students' classrooms than in my own, sharing principles of leadership, **sustainability**, and changemaker skills.

   **a.** the ability to keep going over a long period

   **b.** the ability to operate at low costs

   **c.** the ability to generate a lot of money

2. ... it's OK to take a break and step back for a second. There are many of us on the front lines who will continue the work while you rest and **recharge**.

   **a.** work on something else **b.** get your energy back **c.** do something more than once

3. Something activates you. An experience, an **injustice** that takes place, big or small, local or global. And then there is almost no choice but to get involved.

   **a.** something unfair **b.** something impossible **c.** something unknown

4. And while we wait for the classrooms to adapt, once again, my **peers** and I create our own learning journey.

   **a.** people with different values **b.** people with the same values **c.** people of similar age or position

5. Maybe invite us to one of your board meetings and ask us for some reverse **mentoring** sessions.

   **a.** lessons on how to improve performance **b.** support and advice for less-experienced people **c.** training on effective communication

**E** Use the suffix *-or* to form new words based on the words in the box and complete the sentences below.

| **collaboration** | **mentoring** | **navigate** |
|---|---|---|

1. A ship's ________________ works with maps and charts to identify safe routes for the vessel.
2. To me, a true ________________ is someone who shares my dream and works to achieve it.
3. My ________________ taught me everything I know about the business.

## COMMUNICATE

**F** Work with a partner. Discuss the questions below. Use the words in **bold** and explain your answers.

1. When you work on solving a problem, do you prefer to **collaborate** with others or to work alone?
2. Have you ever had a **mentor** who helped you to get through a new or challenging situation?
3. What do you do to relax and **recharge**?
4. What do you and your **peers** do to help the environment?

# Viewing and Note-taking

LEARNING OBJECTIVES

- Watch and understand a talk about becoming a changemaker
- Notice and use intonation for contrast



## TEDTALKS

**Melati Wijsen** is an Indonesian changemaker who is passionate about climate change. In her TED Talk, *A Roadmap for Young Changemakers,* she talks about how young changemakers can get the skills they need by creating their own learning pathways. She shares advice that can help anybody who wants to make a positive difference in our world.

### BEFORE VIEWING

**A** Read the information about Melati Wijsen. Then listen to three excerpts from her talk. At the end of Excerpts 1 and 2, write a question based on what you think she will say next. Were your questions answered?

**1.** ______________________________

**2.** ______________________________

Environmental changemakers Melati Wijsen (left) and her sister Isabel.

## WHILE VIEWING

**B** **LISTEN FOR MAIN IDEAS** Watch Segment 1 of Wijsen's TED Talk and circle T (true) or F (false).

1. Wijsen felt confident when starting as a changemaker. **T** **F**
2. She thinks it's impossible for real change to start in the classroom. **T** **F**
3. She believes every young person can be a changemaker. **T** **F**

**C** **LISTEN FOR DETAILS** Watch Segment 1 of Wijsen's talk again and complete the concept map.

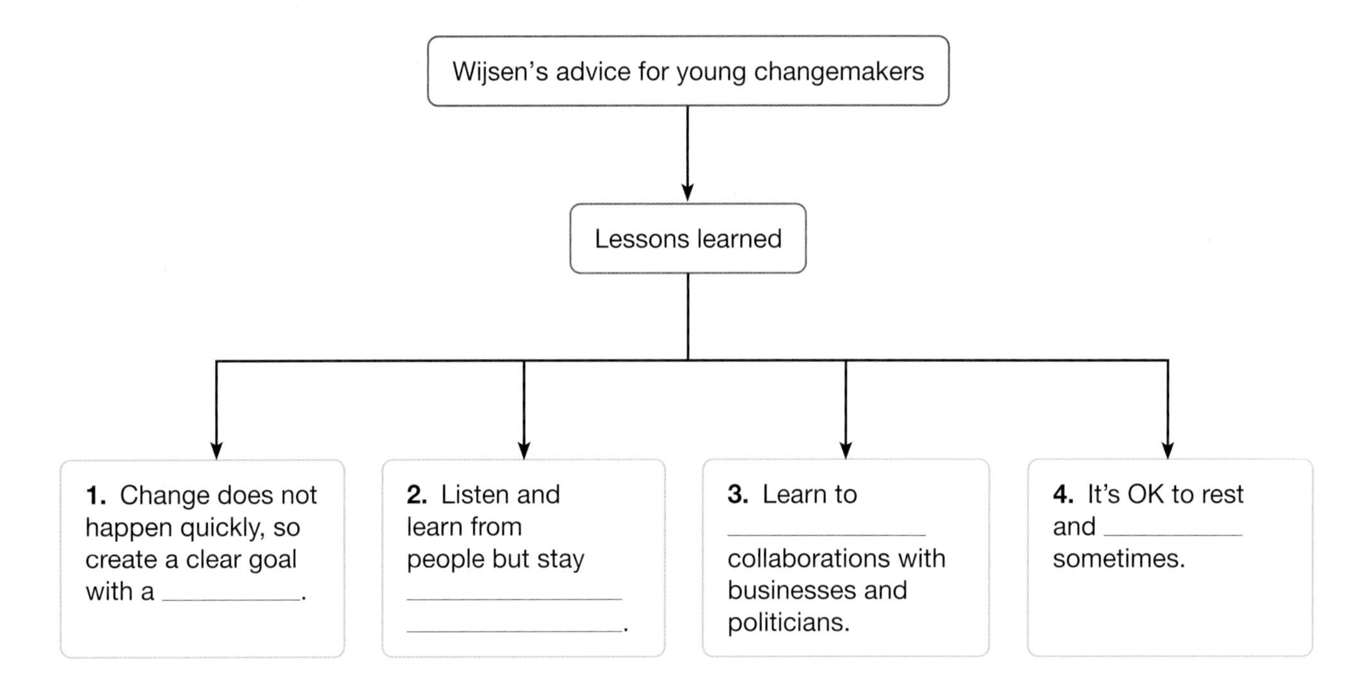


**D** **LISTEN FOR MAIN IDEAS** Watch Segment 2 of Wijsen's talk and select the correct option in each sentence below.

1. This segment is about **greenwashing / youthwashing**.
2. According to Wisjen, some companies want to work with young climate activists because it's good for **business / the planet**.
3. She **would / would not** consider working with companies.

**WORDS IN THE TALK**

*SDGs* (n) the United Nations' Sustainble Development Goals – a set of global goals to work towards a better world.
*mandatory* (adj) required by law

## AFTER VIEWING

**E** Work with a partner. In Wijsen's talk, she says that classrooms often do not reflect reality. What two solutions does she propose to solve this problem?

1. ☐ Students should be role models for one another and learn from one another's experiences.
2. ☐ Students should find a way to help young changemakers who are trying to solve real-world problems.
3. ☐ Students should learn more about the current problems in today's world.
4. ☐ Students should work with people who are affected by climate change to learn more about it.

**F** **ANALYZE** Work with a partner. Do you agree with Wijsen's solutions? What other solutions can you think of? Discuss your ideas.

## PRONUNCIATION *Contrasting intonation*

**G** Look at the example. Notice how ideas are presented in contrast. Then listen to three excerpts from the talk. Underline the ideas she contrasts and add arrows for intonation.

*Listen and be open to learn ↗, but stay true to the mission ↘.*

1. An experience, an injustice that takes place, big or small, local or global.
2. I think it is high time to ensure that what we learn in the classrooms reflect what is happening outside of them.
3. It's the process of conveying a false impression about the climate friendliness of a company product or actions.

> **Pronunciation Skill**
> **Contrasting Intonation**
> To contrast two opposing ideas, use a rising intonation on the first word or phrase and a falling intonation on the second.

**H** Complete the sentences and questions about interests and preferences.

1. I prefer ____________ to ____________.
2. I'd rather ____________ than ____________.
3. Would you rather ____________ or ____________?
4. Which do you like best: ____________ or ____________?

**I** Read statements 1 and 2 to your partner using the correct intonation on the contrasting words or phrases. Explain the reasons for your preferences. Then ask your partner your questions 3 and 4.

# Thinking Critically

LEARNING OBJECTIVES

- Interpret an infographic about youth beliefs and behavior
- Synthesize and evaluate ideas about making a difference

## ANALYZE INFORMATION

**A** Look at the infographic and answer the questions below. Discuss your answers with a partner.

1. What are the two main topics in the survey?
2. Which two statements are true about today's youth, according to the survey?
   - **a.** ☐ They care about inequality and want those with power to do more.
   - **b.** ☐ They felt more positive than negative about the environment.
   - **c.** ☐ About half of them take action by giving money.

## What Do You Believe In?

In 2021, a survey of 8,273 young people born between 1995 and 2003, and coming from 45 different countries, provided some fascinating insights into the beliefs and values of today's youth.

### YOUTH IN 2021

**BELIEFS:** Which of these statements do you agree with?

**66%** Wealth and income are distributed unequally

**62%** Businesses are only interested in making money

**60%** Leaders are not focused on protecting the environment

**43%** It's too late to repair the environmental damage already done

**25%** I'm hopeful about the environment

**16%** Anyone can achieve a high level of wealth

**SOCIAL ENGAGEMENT:** Have you engaged in these activities in the past two years?

**52%** Donated to a cause they support

**49%** Chose work or potential employers based on personal ethics

**40%** Created social media content on an important issue, e.g., the environment

**40%** Been a volunteer or member of a community organization, charity, or nonprofit

**36%** Raised money for charity

Source: The Deloitte Global Millennial And Gen Z Survey, 2021

A volunteer organizes donations in a large food bank.

**B** **LISTEN FOR DETAILS** Listen to part of a lecture about the survey data and complete the information about millennials.

| | Millennials | Gen Z |
|---|---|---|
| **Born between** | **1.** | January 1995 and December 2003 |
| **Number surveyed** | **2.** | 8,273 |
| **Donated to charity** | **3.** | 52% |
| **Volunteered** | **4.** | 40% |
| **Chose work or potential employers based on personal ethics** | **5.** | 49% |

**C** Discuss the questions below with a partner.

1. What information from Exercise A or B was most interesting or surprising to you? Why?
2. Look at the infographic. Which social engagement activities have you done recently? Are there any you would like to do?
3. What can companies do to show they care about people and the environment?

## COMMUNICATE *Synthesize and evaluate ideas*

**D** Based on the survey results in the infographic and Melati Wijsen's TED Talk, choose four issues you think are priorities for young people today and number them 1–4, with 1 being the most important. Note down your reasons.

| | | |
|---|---|---|
| Disease prevention | Anti-bullying | Climate change |
| Equal rights | Consumer rights | Feeding the hungry |
| Protecting the oceans | Wildlife conservation | Education reform |

**E** Work with a partner. Discuss the questions below.

1. Did you choose the same things? Explain your reasons.
2. Are you actively engaged in any of these areas? Would you like to be in the future?

I think that right now equal rights is the number one priority.

Can you say more about why? I chose climate change because that will affect everything ...

# Putting It Together

LEARNING OBJECTIVES

- Research, plan, and present on the topic of making a difference
- Use enthusiasm to engage your audience

**ASSIGNMENT**

**Individual presentation:** You are going to give a presentation about three individuals or organizations that are making an important difference in a cause you believe in.

## PREPARE

**A** Review the unit. What are some ways that individuals or organizations can make a difference in the world? Make a list of ideas that were discussed in the unit.

| Ways to make a difference | |
|---|---|
| **Individuals** | **Organizations** |
| *Do research to learn more about a problem.* | *Have more experienced individuals train those with less experience.* |

**B** Search online for three individuals or organizations that are active for a cause you believe in. Find examples of different actions they are taking to have an impact in the world. Note your findings in the chart below.

| Cause: | | |
|---|---|---|
| **Individual/Organization** | **Action(s)** | **Impact** |
| | | |
| | | |
| | | |

**C** Plan your presentation. Use your notes to help you, and include details and examples.

**D** Look back at the vocabulary, pronunciation, and communication skills you've learned in this unit. What can you use in your presentation? Note any useful language below.

______________________________________________

______________________________________________

______________________________________________

**E** Below are some ways you can show enthusiasm in a presentation. Think about how you can show enthusiasm in your presentation.

- Stress key words
- Add emphasis to key words
- Choose words that have emotional power
- Use facial expressions
- Use gestures

> **Presentation Skill**
> **Showing Enthusiasm**
> In Melati Wijsen's talk, her use of emphasis and her tone of voice help to make her enthusiasm for her topic clear. Often, your enthusiasm will spread to the audience and make them more interested in your ideas.

**F** Practice your presentation. Make use of the presentation skill that you've learned.

## PRESENT

**G** Give your presentation to a partner. Watch their presentation and evaluate them using the Presentation Scoring Rubrics at the back of the book.

**H** Discuss your evaluation with your partner. Give feedback on two things they did well and two areas for improvement.

# Checkpoint

Reflect on what you have learned. Check your progress.

**I can ...** ☐ understand and use words related to equality.

| | | | | |
|---|---|---|---|---|
| **collaboration** | **convey** | **essential** | **extend** | **injustice** |
| **mentoring** | **navigate** | **peer** | **recharge** | **sustainability** |

☐ use the suffix *-or.*

☐ watch and understand a talk about becoming a changemaker.

☐ notice and use intonation for contrast.

☐ interpret an infographic about youth beliefs and behavior.

☐ synthesize and evaluate ideas about making a difference.

☐ use enthusiasm to engage an audience.

☐ give a presentation on individuals or organizations making a difference in the world.

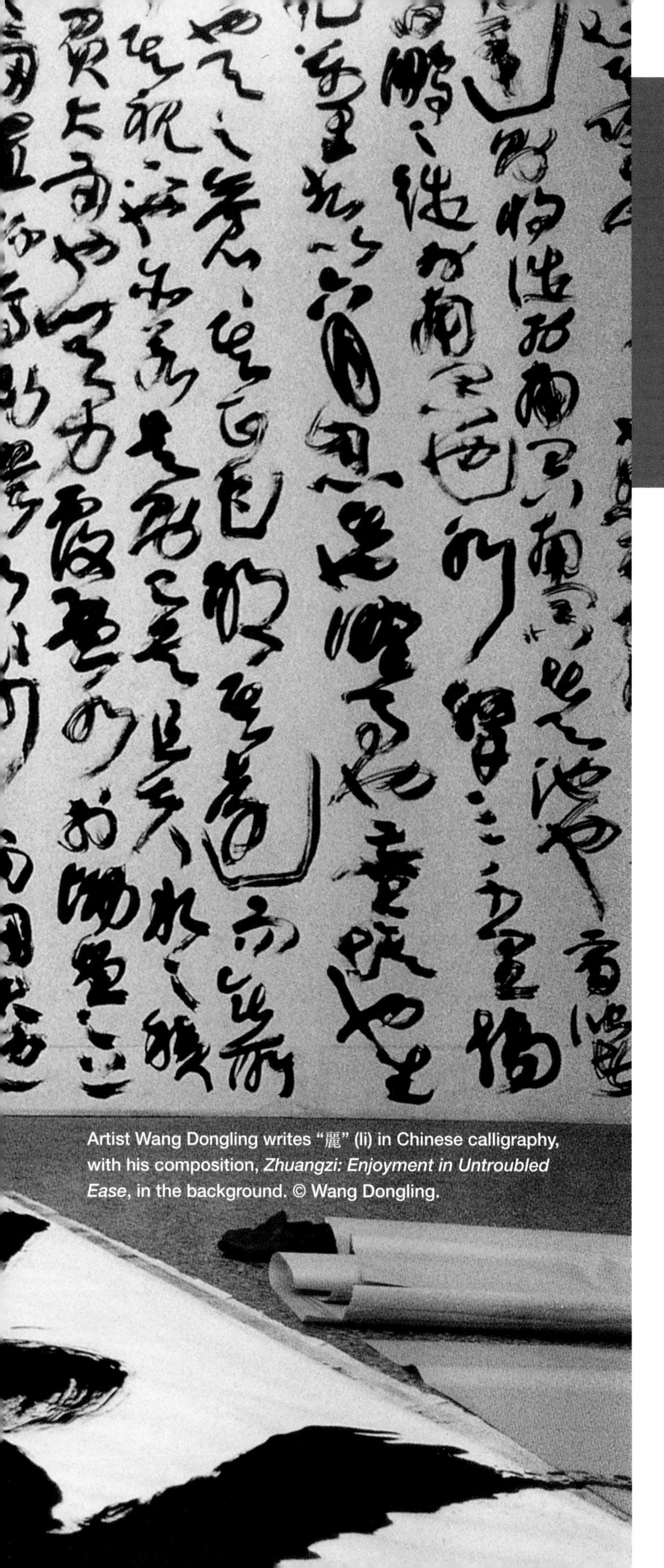

Artist Wang Dongling writes "麗" (li) in Chinese calligraphy, with his composition, *Zhuangzi: Enjoyment in Untroubled Ease*, in the background. © Wang Dongling.

# 3

# Say It Your Way

## Q How can we become better communicators?

Artists like Wang Dongling (pictured) use their art as a way to communicate ideas, thoughts, and feelings with people around the world. Art has been used this way for thousands of years, but this is just one of many modes of communication that we have developed over time. Smartphones are one of our most recent creations—but how are they affecting our communication skills? In this unit, we explore this, and look at other ways that we can communicate new ideas effectively.

## THINK and DISCUSS

1 Look at the photo and the unit title. What ideas do you think the artist might be trying to share with us?

2 Look at the essential question and the unit introduction. Have you ever had trouble making yourself understood by other people? Why?

UNIT 3

# Building Vocabulary

LEARNING OBJECTIVES

- Use ten words related to communication
- Use suffixes to change word forms

## LEARN KEY WORDS

**A** Listen to and read the information in the timeline. Which milestone in the history of human communication do you think was most important? Discuss with a partner.

**B** Match the correct form of each word in **bold** from Exercise A with its meaning.

1. ______________ a shortened form of a phrase
2. ______________ to communicate a message, meaning, or idea
3. ______________ a family member who lived a very long time ago
4. ______________ a body movement to show something, such as a feeling
5. ______________ not in a straight course or path
6. ______________ not simple; complicated
7. ______________ found or practiced everywhere
8. ______________ a look on someone's face that shows how they feel
9. ______________ a sign, mark, picture, or object that represents something else
10. ______________ something designed and created that has never been made before

# History of Communication

Before humans began to use language, our **ancestors** communicated through **facial expressions** and **gestures**. Since then, human communication has evolved. Through speech, writing, and images, people fulfill the **universal** need to communicate with one another.

**PICTOGRAPH WRITING SYSTEMS — HIEROGLYPHICS:** Pictographs became more **complex**, forming the world's first writing systems

**62,000 BCE**

**CAVE PAINTINGS:** Some of the oldest cave paintings in the world are in Chauvet Cave, France. They show a variety of different animals, including woolly mammoths.

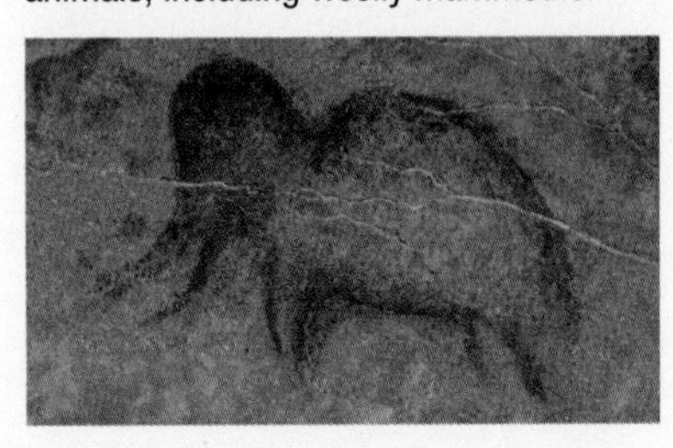

**5,000 BCE**

**PICTOGRAPHS:** These **symbols** were used to represent an object or event. They began to appear around the world about 7,000 years ago.

**3,300 BCE**

**800 CE**

**WOODBLOCK PRINTING:** Woodblock printing originated in China and was used on textiles and paper.

**C** Use the suffixes in the box to complete the table below with the different word forms. Combine two suffixes if necessary. Use a dictionary if necessary.

| -ity | -ic | -ize | -ly | -al | -ness | -tion |
|---|---|---|---|---|---|---|

| Adjective | Noun | Verb | Adverb |
|---|---|---|---|
| complex | | X | X |
| | symbol | | |
| | universe | X | |
| indirect | | X | |
| reduced | | reduce | X |

**D** Complete the sentences below with the correct form of the words in the box.

| gesture | invent | ancestor | facial expression |
|---|---|---|---|

1. No one can be sure why our ______________ made paintings of animals in caves.
2. Modern technology means that ideas can rapidly become new ______________.
3. On video calls you can see ______________ and hand ______________.

## COMMUNICATE

**E** Discuss with a partner. What kinds of communication methods do you use most often, and with whom? Have these methods changed much over time?

**PHOTOGRAPHY:** The first photo portrait was a selfie by Robert Cornelius in 1839.

**EMAIL:** The first email was sent by Ray Tomlinson to himself.

**1450 CE** **1826 CE** **1876 CE** **1971 CE** **1992 CE**

**PRINTING PRESS:** With the **invention** of Gutenberg's printing press, large amounts of printed material began to be produced. Ideas could spread across large distances.

**TELEPHONE:** Scottish-born inventor Alexander Graham Bell made the first ever telephone call to his colleague Thomas Watson.

**MESSAGING:** In December 1992, British engineer Neil Papworth sent the world's first SMS. **Reductions** are often used in phone texts to **get** messages **across** while saving valuable time and space. Many people today have a preference for more **indirect** communication and choose texting over talking on the phone.

MERRY CHRISTMAS

The world's first SMS

# Viewing and Note-taking

LEARNING OBJECTIVES

- Watch a video podcast about emojis and emoticons
- Include only essential details when taking notes
- Listen for explanations of words and terms

## BEFORE VIEWING

**A** Read the types of information that people usually focus on in their notes. Match them with the examples. What other types of information could you add to the list? Discuss with a partner.

**1.** _______ numbers and statistics

**2.** _______ definitions of key vocabulary

**3.** _______ one example of something you find difficult to understand

**4.** _______ the reasons for something

**5.** _______ explanations of how things work

**6.** _______ the steps in a process

**a.** *#1: Warm up the oven by setting the temperature to ...*

**b.** *55% students believe that ...*

**c.** *causes of extreme weather: bad luck, global warming ...*

**d.** *squish = to squeeze something*

**e.** *lungs expel air → vocal cords vibrate → muscles and tissues in the ...*

**f.** *discrimination, e.g., harassment, not having fair rights to ...*

**B** Listen to these extracts from a podcast. Take notes on the essential details. Compare your notes with a partner.

**1.** Hearing loss and deafness

______________________________

______________________________

______________________________

______________________________

**2.** Sign language

______________________________

______________________________

______________________________

______________________________

**Note-taking Skill**

**Including Only Essential Details**

When you take notes, include only the details that seem essential to help you understand and remember the speaker's main points. Write short phrases and use abbreviations to save time. It may be useful to draw a sketch if the speaker is describing something visual.

**C** Work with a partner. You are going to watch a video podcast called *Communicating with Emojis and Emoticons.* Circle what you think the podcast will be about.

**1.** Which is better: emojis or emoticons?

**2.** Popular and unpopular emojis and emoticons

**3.** The pros and cons of using emojis and emoticons

## WHILE VIEWING

**D** ▶ **LISTEN FOR MAIN IDEAS** Watch Segment 1 of the video podcast and check (✓) two main ideas that the speaker shares.

1. ☐ Emojis and emoticons are less direct than written words.
2. ☐ Emojis and emoticons can be used to communicate with other people.
3. ☐ Emojis and emoticons express universal human feelings.
4. ☐ Emojis are more useful than emoticons because they are used all over the world.

**E** ▶ **LISTEN FOR EXPLANATIONS** Watch Segment 1 of the podcast again and answer the questions below. Discuss your answers with a partner.

1. What are emoticons?
2. What are emoticons made from?
3. What is a compound word?
4. What does the *e* in emoji mean?
5. What does *moji* mean?

> **Listening Skill**
>
> **Listening for Explanations**
>
> A speaker may provide explanations of content-specific words and terms. Listening for those will help you better understand the new vocabulary and the speaker's ideas.

**F** ▶ **LISTEN FOR MAIN IDEAS** Watch Segment 2 of the podcast and number the topics 1–3 in the order the speaker talks about them.

**a.** ______ arguments in favor of emojis and emoticons

**b.** ______ the connection between texting, writing, and speaking

**c.** ______ arguments against emojis and emoticons

**G** ▶ **LISTEN FOR DETAILS** Watch Segment 2 of the podcast again and note down the essential details relating to each topic in Exercise F. Discuss with a partner.

______________________________________________

______________________________________________

______________________________________________

______________________________________________

## AFTER VIEWING

**H** **APPLY** Work in a group. Can you guess the meaning of each pair of emojis below? They are all compound words.

1. You write in it
2. Wedding holiday
3. Part of a website

# Noticing Language

LEARNING OBJECTIVES

- Notice language for explanations
- Explain the meaning of unfamiliar terms

## LISTEN FOR LANGUAGE *Explain unfamiliar terms*

**A** Read the different ways to explain unfamiliar terms below. Why do we need these different ways? Discuss with a partner.

**Ways to explain an unfamiliar term:**

| | |
|---|---|
| A definition: | What a term means |
| An analogy: | How a term is similar to something else |
| A description: | Further details about a term's function or purpose |
| An origin: | How a term was formed |
| An example: | One way that a term is used |

**Communication Skill**

**Explaining Unfamiliar Terms**

When you explain specific words and terms that may be unfamiliar to your listeners, you can use different types of explanations to make the meaning clear.

**B** Listen to excerpts from the podcast in Lesson B. What type of explanation did the speaker provide in each case? Discuss with a partner.

| definition | analogy | description | origin | example |
|---|---|---|---|---|

**1.** ____________ **3.** ____________ **5.** ____________

**2.** ____________ **4.** ____________

**C** Look at the terms below and write an explanation for each one. Try to use a different type of explanation for each term.

**E.g. Ancestors** family members who lived a very long time ago (definition)

**1.** Invention ____________

**2.** Facial expression ____________

**3.** Gesture ____________

**4.** Symbol ____________

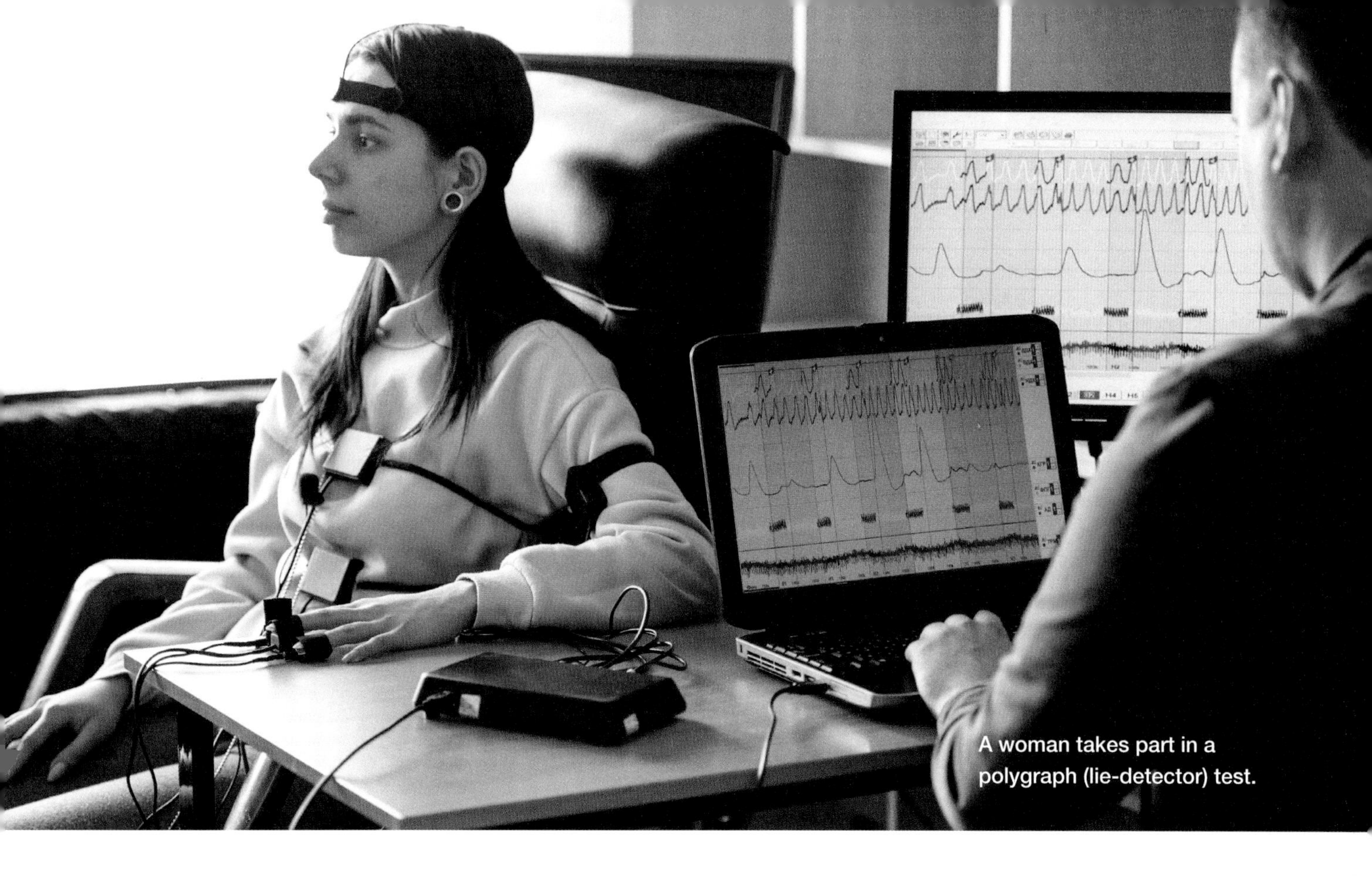
A woman takes part in a polygraph (lie-detector) test.

**D** You are going to listen to a short conversation between a professor and a student about the body language people use when they lie. How do you think the professor might explain the terms below? Discuss with a partner.

| Term | Explanation type(s) |
| --- | --- |
| Fidget | ______________________ |
| Defense mechanism | ______________________ |

**E** Listen to the conversation. What kind of explanations does the professor use to get her meaning across?

## COMMUNICATE

**F** Work with a partner. Take turns reading your explanations from Exercise C in a random order. Can you guess each other's words?

**G** Work with a partner. Take turns to explain the words below. Use at least two of the explanation types each time.

**1.** Woolly mammoth **2.** Hieroglyphic **3.** Philosophy **4.** Thesaurus

**H** Search online to find some interesting compound words. Take turns explaining them to a partner, and try to guess each other's words.

# Communicating Ideas

LEARNING OBJECTIVES

- Use appropriate language to explain new terms
- Collaborate to explain two new terms

**ASSIGNMENT**

**Task:** You are going to collaborate with a partner to create two new words and teach them to a partner.

## READ FOR INFORMATION

**A** **READ FOR MAIN IDEAS** Read the passage. What is the purpose of the text? Where might you see a text like this? Discuss your ideas with a partner.

**Calling for submissions for our Online Dictionary!**

Our team at JX Lexicographers are working on a new edition of our Online Dictionary. Language never stops changing and we are changing along with it! We're currently looking for a special type of compound word called a blend. In blend words at least one of the joined words is shortened in some way. For example, when you can't decide if someone's a friend or an enemy, they are your *frenemy*. And what about when you're so hungry that it makes you *hangry*?

We're looking for blend words in these categories:

- Food
- Hobbies
- School and studies
- Friends and family
- Technology

To participate, click here to submit your new word on our "Competitions" page. Vote for your favorite entry by March 31, and the ten most popular words will be included in our dictionary!

**B** Use the chart below to help you note the essential details of the passage above.

| Competition: JX Lexicographers looking for new blend words | |
|---|---|
| Definition(s) of a blend: | Examples(s) of blend words: |
| How to participate: | Reason to participate: |

## COLLABORATE

**C** Work with a partner. Choose one of the categories listed in the passage. Brainstorm a list of words related to each one, for example, types of activities, people, things, and emotions.

**category:** ______________________________

**words:** ______________________________

______________________________

**D** With your partner, create new words by blending two words in your category. Come up with at least three ideas.

______________________________

______________________________

**E** Select your two favourite new words and prepare to explain them. For each word, write a definition and include another form of explanation (e.g., analogy, further description, example).

______________________________

______________________________

**F** Work with a different partner to share your two new words. Follow these steps:

1. Present your two new words to your partner.
2. Let your partner try to guess the words. Tell them if they guessed correctly.
3. Provide your explanation of the meaning of the two new words.

# Checkpoint

Reflect on what you have learned. Check your progress.

**I can ...** understand and use words related to communication.

| | | | | |
|---|---|---|---|---|
| **ancestor** | **complex** | **facial expression** | **gesture** | **get across** |
| **indirect** | **invention** | **reduction** | **symbol** | **universal** |

use suffixes to change word forms.

watch and understand a podcast about emojis and emoticons.

focus on essential details when taking notes.

listen for explanations of words and terms.

notice language to explain new terms.

use language to explain the meaning of unfamiliar terms.

collaborate and communicate effectively to create and explain two new terms.

The “Love Wall” in Paris, France, is covered with “I love you” in over 300 languages.

# Building Vocabulary

LEARNING OBJECTIVES

- Use ten words related to language
- Understand collocations with *attention*

## LEARN KEY WORDS

**A** Listen to and read the passage below. What are "fillers" and why do people use them? Discuss with a partner.

**Filling the Gaps**

"Umm ...," "Actually, ...," "OK, ..." English speakers tend to use a lot of short words and phrases like these in conversation, often without realizing it. **Linguists** refer to them as discourse markers, but they are more commonly known as "fillers"; that is, they help to fill the silence as you think of what to say. Although **usage** of these small words is often **unconscious** on the part of the speaker, they play an important role in helping us communicate.

Fillers have a variety of functions. Some help us soften a statement: "*Well*, I'm not sure about that ...". Others help us take the lead in a conversation so we can **grab** our listener's **attention**: "*Hey*, I have an idea ...," "*Listen*, how about we ...?" Meaning can also **shift** depending on how a filler is used. "Mmm ..." for example can indicate someone is interested in what you are saying, but it can also express boredom or disbelief—it all depends on the speaker's intonation.

**B** Work with a partner. Discuss the questions below.

1. Can you think of any other filler words in English? How does each one affect what you say?
2. Look at the photo. Do you recognize any of the languages on the Love Wall? What languages can *you* say "I love you" in?

**C** Circle the best definition of the word in **bold** in each sentence.

1. Word **usage** changes over time. For example, before Facebook existed, *friend* was only used as a noun. However, many people now also use it as a verb.

   **a.** how something is defined **b.** how popular something is **c.** how something is used

2. People who write dictionaries are usually **linguists**.

   **a.** interested in language **b.** language experts **c.** language teachers

3. Our use of our native language is **unconscious**. We may not be aware of the grammar rules, but we can still speak correctly.

   **a.** difficult to explain **b.** understandable **c.** in our mind but unknown to us

4. The meanings of words can change over time. For example, in the 14th century, *girl* meant a female or male child. Later, there was a **shift** in meaning, and *girl* now only refers to females.

   **a.** change **b.** difference **c.** problem

5. Unusual words **grab** people's **attention**. When you use them, people listen carefully to you.

   **a.** attract interest **b.** interfere with understanding **c.** improve understanding

**D** Read the excerpts from Erin McKean's TED Talk in Lesson F. Then choose the options that are closest to the meaning of the words in **bold**.

1. "So another way that you can make words in English is by squishing two other English words together. This is called compounding."

   When you **compound** things, you ____________.

   **a.** reduce them **b.** crush them **c.** join them

2. "'Commercial' used to be an adjective and now it's a noun."

   Something that is **commercial** is something that is mainly related to ____________.

   **a.** art **b.** selling **c.** justice

3. "Now, sometimes people use this kind of rules-based grammar to discourage people from making up words."

   When you **discourage** someone, you want to ____________ them from doing something.

   **a.** prevent **b.** reward **c.** motivate

4. "You use so much force when you squish the words together that some parts fall off. So these are **blend** words ..."

   When you **blend** two things, you ____________.

   **a.** confuse them **b.** destroy them **c.** mix them

5. "Now, there are other rules that are more about manners than they are about nature."

   If someone has good **manners**, they ____________.

   **a.** behave well **b.** dress well **c.** speak well

**E** Many verbs collocate with the word *attention*. Complete the sentences using the correct form of the collocations.

| pay attention | give attention | draw attention | get attention | hold attention |
|---|---|---|---|---|

1. Now, I'd like to ____________ your ____________ to the next slide of my talk.
2. The speaker yesterday was fantastic—she ____________ our ____________ for hours.
3. This is an important announcement, so please ____________.
4. They called his name loudly to ____________ his ____________.
5. I don't know why she bought a pet. She never ____________ it any ____________.

## COMMUNICATE

**F** Work with a partner. Discuss to what extent you agree with the statements below.

1. Teachers should **discourage** their students from using slang.
2. It's bad **manners** to interrupt people when they are speaking.
3. Children who behave badly are just trying to **get attention**.

# Viewing and Note-taking

LEARNING OBJECTIVES

- Watch and understand a talk about creating new words
- Notice the use of stress in compound words



## TEDTALKS

Lexicographer **Erin McKean** created Wordnik, an online dictionary that contains not only traditionally accepted words and definitions, but also new words and new uses for old words. She is also an active blogger and the author of many books. In her TED Talk, *Go Ahead, Make Up New Words*, she explains how making up new words will help us use language to express what we mean.

### BEFORE VIEWING

**A** Read the information about Erin McKean. Discuss the questions below with a partner.

1. Have you ever made up any new words, either in English or in another language you speak?
2. Do new words appear in your language(s) every year? If so, where do you think they come from? Who do you think creates them?
3. What are some examples of new words in your language(s)? Why do you think those words appeared?

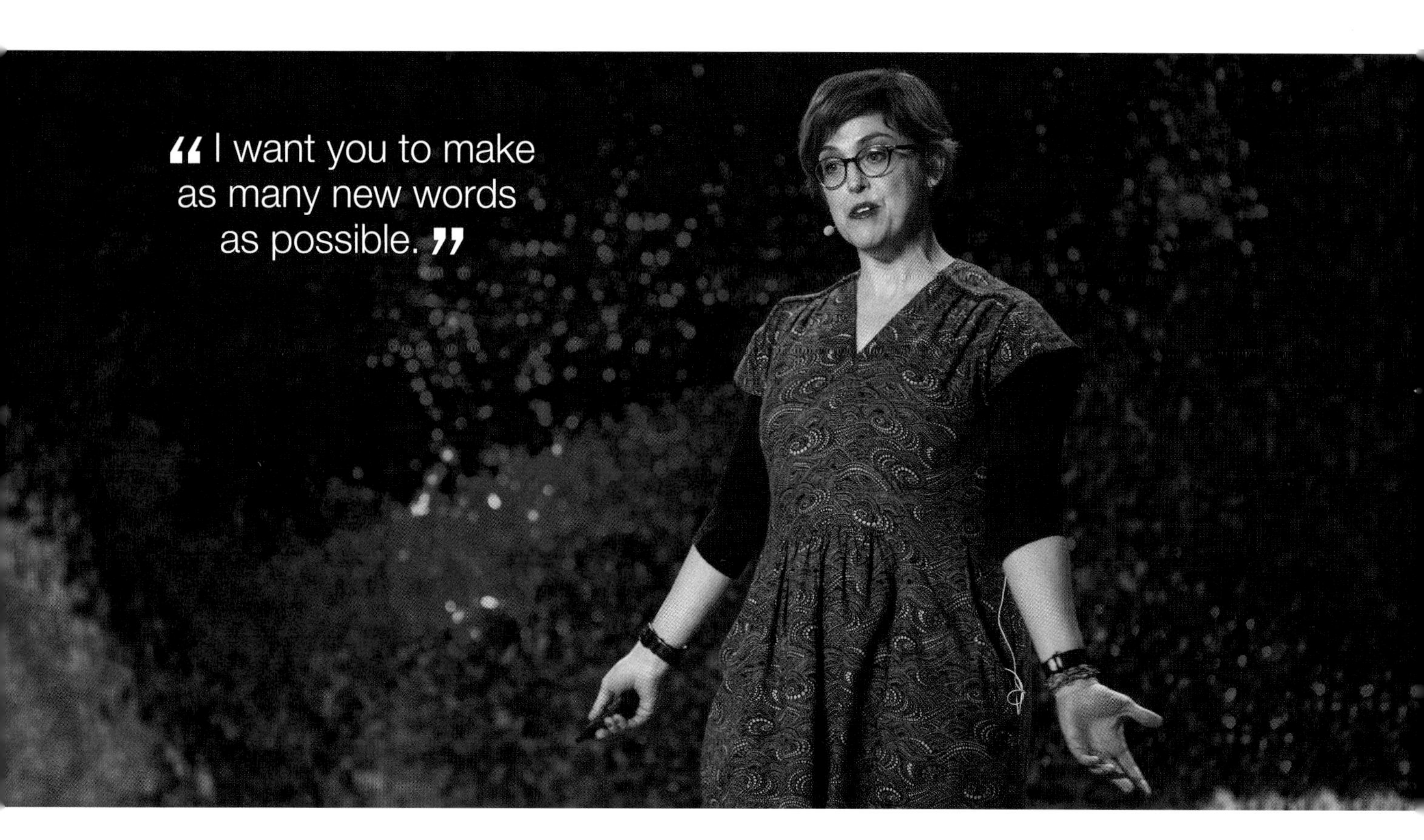

## WHILE VIEWING

**B** ▶ **LISTEN FOR MAIN IDEAS** Watch Segment 1 of Erin McKean's TED Talk and number the topics 1–3 in the order she talks about them.

**a.** ____ One kind of grammar uses rules that we have to be taught.

**b.** ____ A focus on grammar can discourage people from making up words.

**c.** ____ One kind of grammar uses rules that we follow unconsciously.

**C** ▶ **LISTEN FOR DETAILS** Watch Segment 2 of McKean's TED Talk and take notes in the graphic organizer below.

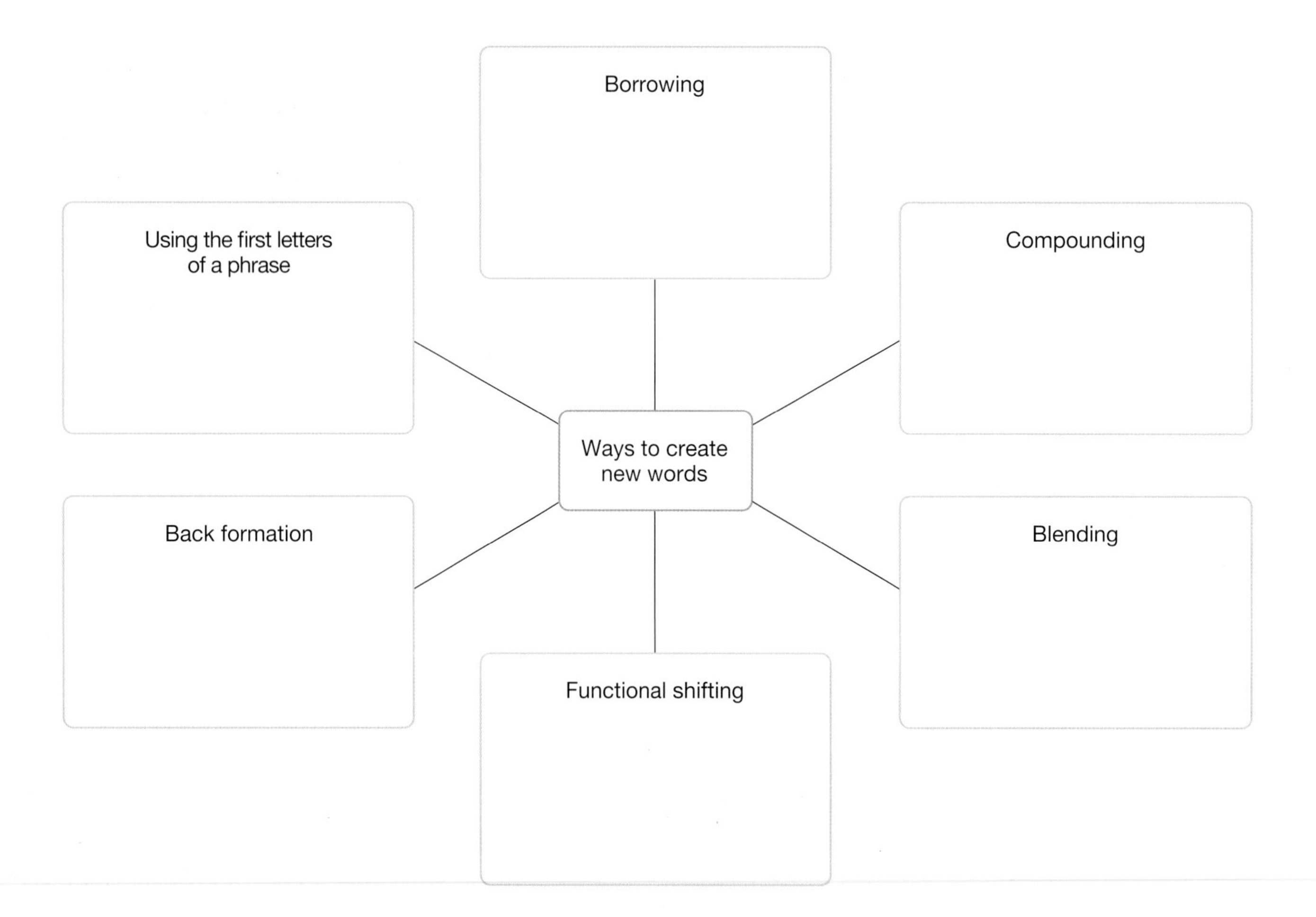


**D** Look back at Exercise A in Lesson B and review the types of information that we usually focus on in notes. Which types of information did you include in your notes above? Did you miss anything important? Discuss with a partner.

**WORDS IN THE TALK**
*duckface* (n) someone's face when they push their lips outward, usually in photos

## AFTER VIEWING

**E** **APPLY** Work with a partner. Take turns explaining the meanings of the words in the box. Then match them with numbers 1–5 according to how they were formed.

| boutique | camcorder | to parent someone | LOL | thunderstorm |
|---|---|---|---|---|

1. ________________ borrowing
2. ________________ compounding
3. ________________ blending
4. ________________ functional shift
5. ________________ using initial letters to form a new word



## PRONUNCIATION *Compound word stress*

**F** Listen to the examples of compound words from the talk. Practice reading the sentence aloud.

*Words like "heartbroken," "bookworm," "sandcastle" all are compounds.*

> **Pronunciation Skill**
> **Stress in Compound Words**
> When you pronounce a compound word (a combination of two words to make one word), the stress is almost always on the first word in the compound.

**G** Find and underline the seven compound words in the statements below. Practice reading the sentences aloud and saying the compound words correctly.

1. When learning a language, getting positive feedback from teachers and classmates is necessary.
2. The most important function of smartphones is to allow people to visit websites and use social media anywhere.
3. Email has helped everyone to communicate effectively in the workplace.

**H** Work with a partner. Do you agree or disagree with the statements in Exercise G? Give reasons or examples to support your position.

The word "sandcastle" is an example of a compound word. Compounds can be one word, two words, or hyphenated.

# Thinking Critically

LEARNING OBJECTIVES

- Interpret an infographic about how a new word enters the dictionary
- Synthesize and evaluate ideas about the acceptance of new words

## ANALYZE INFORMATION

**A** Look at the infographic and answer the questions. Discuss your answers with a partner.

1. Who or what is involved in finding new words to add to a dictionary?
2. What are the requirements for a word to be included in the dictionary? Why are these important?
3. Who is in charge of making sure new words meet all three of the requirements?

## HOW A NEW WORD GETS INTO A TRADITIONAL DICTIONARY

People make up new words all the time. But only some words successfully make the journey from being created to being entered into a dictionary. What are the steps along the way?

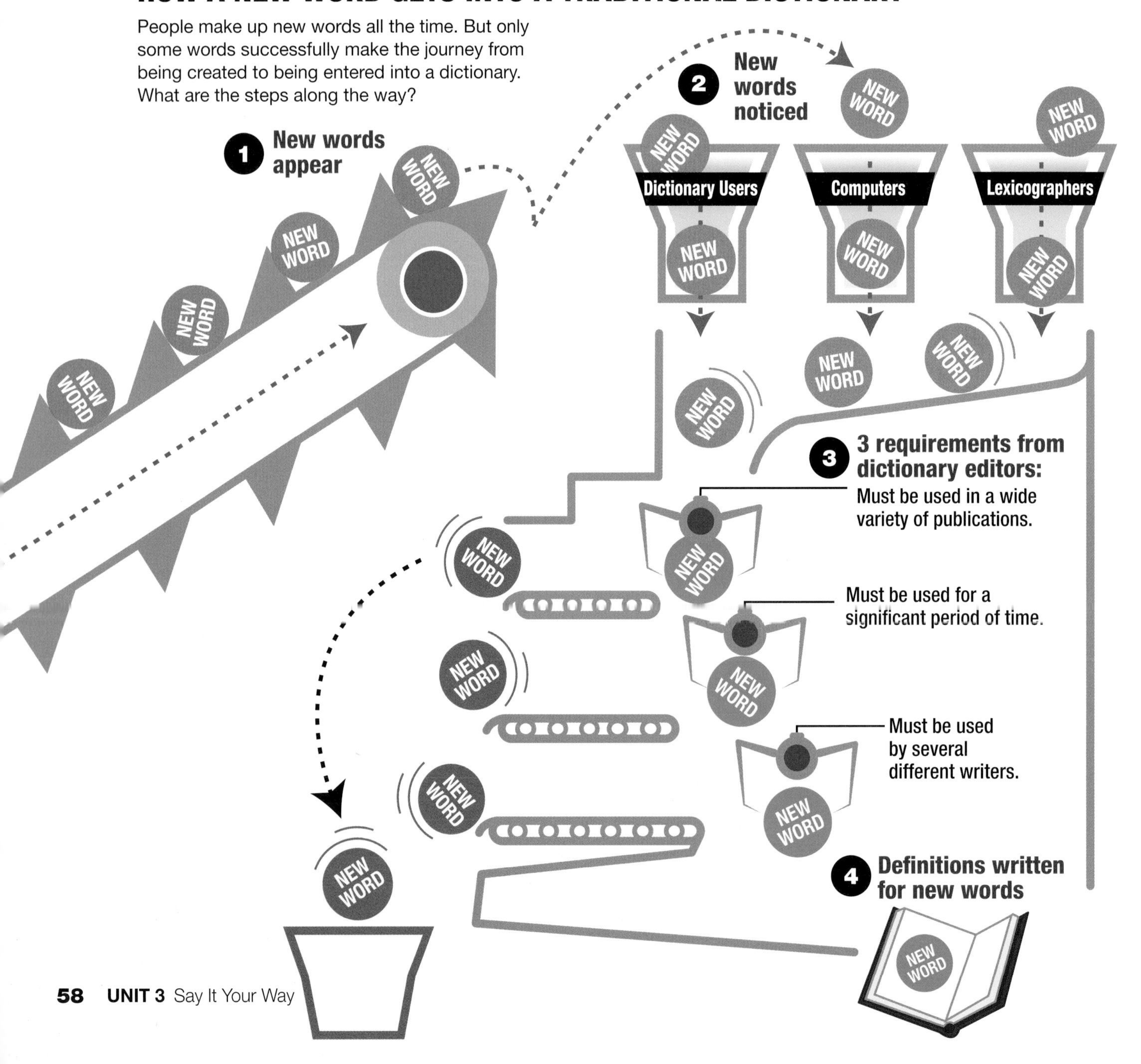

**B** Work with a partner. Read the magazine extract below about a new word, *hygge*. Do you think you will find this word in a traditional dictionary? Discuss your ideas.

It's a cold winter's evening. You're at home in a warm and cozy living room, sipping a coffee and enjoying a pastry while you chat with your friends. That's a very hygge thing to do. Hygge, pronounced *hoogah*, is a word of Danish origin. While it has been part of Danish culture for decades to describe feelings of being comfortable and content, it is only in the last few years that the word has appeared in English. It first appeared in lifestyle books and magazines but has since become more commonly used in TV shows, movies, and social media.

## COMMUNICATE *Synthesize and evaluate ideas*

**C** Take notes on the questions below. Then discuss your ideas with a partner.

1. How long do you think the process of adding a word to a dictionary took in the past? Days? Months? Years? How long does it take today?
2. How do you think the process Erin McKean uses for her own online dictionary, Wordnik, might be different from the process shown in the infographic?
3. Why do you think McKean decided to start her own online dictionary?

**D** Work with a partner. Here are some words from other languages that are sometimes used in English. Have you heard them before? Do they appear in your dictionary? Why, or why not, do you think?

*kawaii* (adj) – very cute (Japanese)

*bibimbap* (n) – rice dish with vegetables or meat and an egg on top (Korean)

*faux pas* (n) – a mistake in a social situation (French)

*Schadenfreude* (n) – enjoyment of someone else's misfortune (German)

**E** Work in a small group. Which words from your language(s) do you think English should borrow? Use appropriate ways to explain the meaning clearly.

In Japan, we have the word "mottainai". It's about not wasting anything. It's "mottainai" to throw away vegetable scraps when you can use them to make soup. It's an important word in the environmental movement.

# Putting It Together

LEARNING OBJECTIVES

- Research, plan, and present on the topic of uncommon words.
- Use language to encourage audience participation

**ASSIGNMENT**

**Group presentation:** Your group is going to research uncommon words and give a presentation to encourage your audience to use three of the most useful ones.

## PREPARE

**A** Review the unit. What ideas have you learned about how to get your meaning across more effectively? Discuss with a partner and take notes.

**B** Work with your group. Search online for uncommon words. Look at the categories below. What unusual words can you find in these categories? Use the chart to take notes on the words and their definitions.

| **Emotions** | **Characteristics** |
|---|---|
| **People** | **Animals** |
| **Things** | **Other** |

**C** Plan your presentation with your group. Follow the steps below.

1. Decide on the three words that you think are most interesting or useful.
2. Decide on the best way to explain the words. Can you support definitions with analogies or examples?
3. Look back at Exercise A and think about ways to help your classmates understand your explanations better. What can you add to your presentation? Do you need to provide visuals when presenting?
4. Decide who will give each part of the presentation, ensuring everyone takes a turn.

**D** Look back at the vocabulary, pronunciation, and communication skills you've learned in this unit. What can you use in your part of the presentation? Note any useful language.

**E** Below are some ways to engage an audience. Think about how you can use these in your presentation.

- Be relaxed and friendly—smile!
- Ask the audience to do something, e.g., "Raise your hand if ..."
- Ask the audience questions.
- Ask the audience follow-up questions if no one answers.
- Respond encouragingly even when answers are incorrect.
- At the end of the presentation, thank your audience for listening and participating.

> **Presentation Skill**
>
> **Encouraging Audience Participation**
>
> In an interactive presentation, like Erin McKean's, the speaker encourages the audience to participate from time to time. An engaged audience is more likely to pay attention to your presentation and to embrace your ideas.

**F** Practice your presentation. Make use of the presentation skill that you've learned and some of the useful language from the box below.

> **Language to encourage participation**
> - How about you? Do you want to give it a try?
> - Let me give you a hint. The first part means .... Can you guess what the second part means?
> - Great, thank you. Any other guesses?

## PRESENT

**G** Give your presentation to another group. Watch their presentation and evaluate them using the Presentation Scoring Rubrics at the back of the book.

**H** Discuss your evaluation with the other group. Give feedback on two things they did well and two areas for improvement.

## Checkpoint

Reflect on what you have learned. Check your progress.

**I can ...**

☐ understand and use words and phrases related to language.

| | | | | |
|---|---|---|---|---|
| **blend** | **commercial** | **compound** | **discourage** | **grab attention** |
| **linguist** | **manners** | **shift** | **unconscious** | **usage** |

☐ use collocations with *attention*.
☐ watch and understand a talk about creating new words.
☐ notice the use of stress in compound words.
☐ interpret an infographic about how a new word enters the dictionary.
☐ synthesize and evaluate ideas about the creation of new words.
☐ use techniques to encourage audience participation.
☐ give a presentation on the topic of uncommon words.

A ballet dancer waits to go on stage during a performance of *The Nutcracker* in Leeds, England.

# 4

# Stress: Friend or Foe?

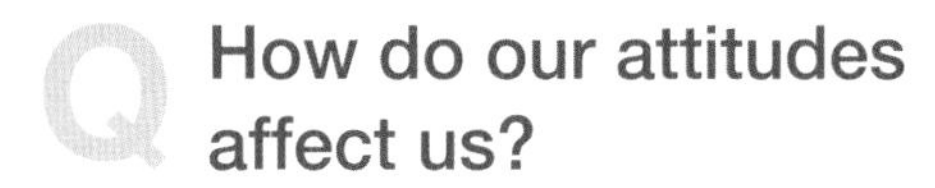

## How do our attitudes affect us?

The dancer in this photo is about to go on stage in front of hundreds of people. For many people, this would be a stressful experience, but not everyone reacts to stress in the same way. And is stress always a bad thing? In this unit, we'll look at what stress and anxiety really are, and strategies that can help us through stressful situations.

## THINK and DISCUSS

1 Look at the photo and read the caption. How do you think the dancer is feeling?

2 Look at the essential question and the unit introduction. What kind of stresses do people face today?

UNIT 4

A B C D

# Building Vocabulary

LEARNING OBJECTIVES

- Use ten words related to stress and anxiety
- Use collocations with words related to stress

## LEARN KEY WORDS

**A** Listen to and read the information below. Do you often experience feelings of stress or anxiety? How do you deal with these feelings? Discuss with a partner.

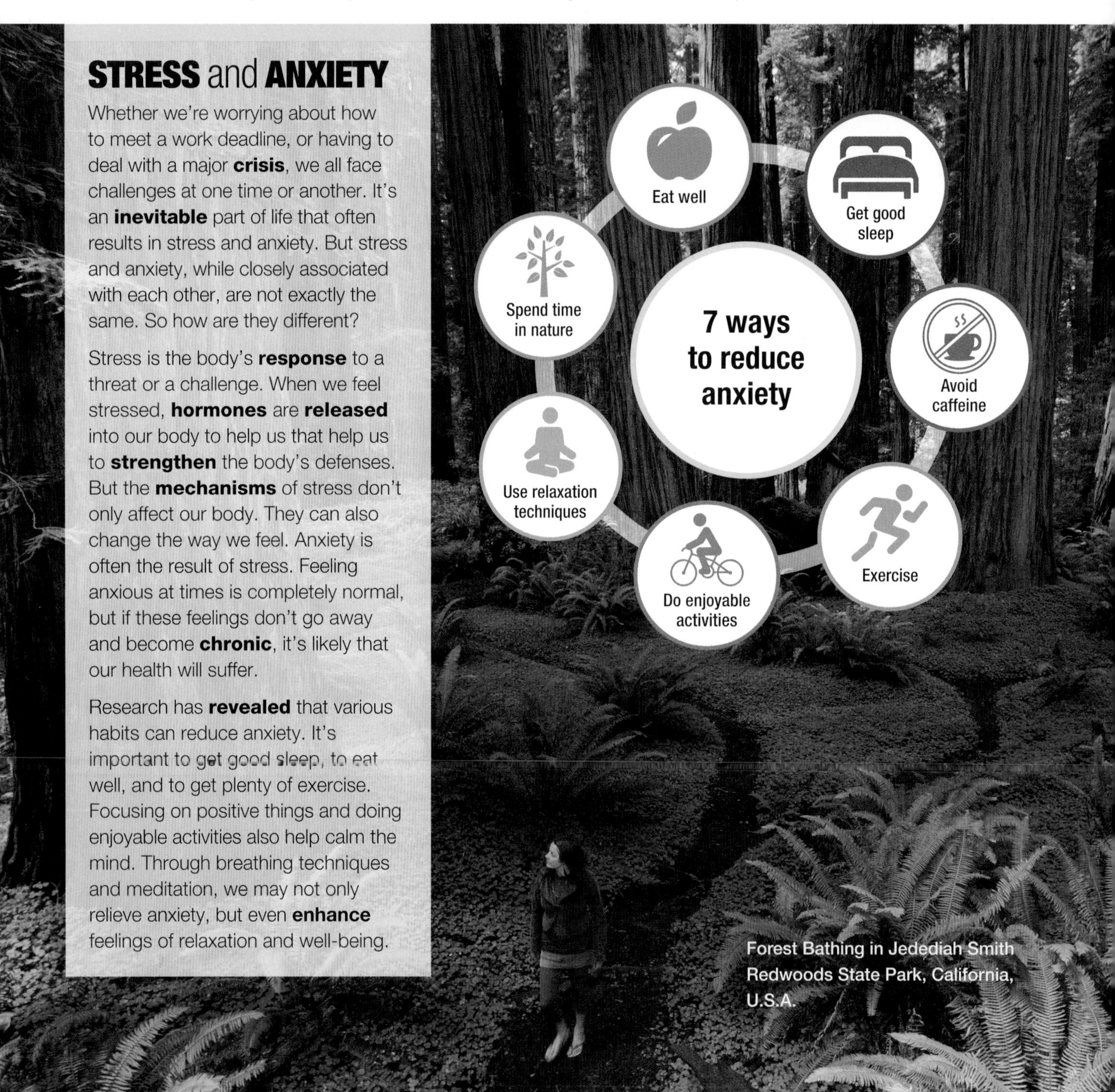


### STRESS and ANXIETY

Whether we're worrying about how to meet a work deadline, or having to deal with a major **crisis**, we all face challenges at one time or another. It's an **inevitable** part of life that often results in stress and anxiety. But stress and anxiety, while closely associated with each other, are not exactly the same. So how are they different?

Stress is the body's **response** to a threat or a challenge. When we feel stressed, **hormones** are **released** into our body to help us that help us to **strengthen** the body's defenses. But the **mechanisms** of stress don't only affect our body. They can also change the way we feel. Anxiety is often the result of stress. Feeling anxious at times is completely normal, but if these feelings don't go away and become **chronic**, it's likely that our health will suffer.

Research has **revealed** that various habits can reduce anxiety. It's important to get good sleep, to eat well, and to get plenty of exercise. Focusing on positive things and doing enjoyable activities also help calm the mind. Through breathing techniques and meditation, we may not only relieve anxiety, but even **enhance** feelings of relaxation and well-being.

Forest Bathing in Jedediah Smith Redwoods State Park, California, U.S.A.

**B** Match the correct form of each word in **bold** in Exercise A with its meaning.

1. ____________ a reaction to something
2. ____________ unavoidable
3. ____________ a system or process
4. ____________ to let go, allow out
5. ____________ to make stronger
6. ____________ to make known, to show
7. ____________ to (further) improve something
8. ____________ an emergency
9. ____________ a chemical that affects our body
10. ____________ ongoing, or frequently returning

**C** The words in **bold** in the sentences below collocate with the words in the box. Complete the sentences with the correct form of the words in the box.

| chronic | hormone | mechanism | enhance | reveal |
|---|---|---|---|---|

1. The **findings** ____________ that the technique was effective at reducing anxiety.
2. Our body ____________ often **work** without us being consciously aware of them.
3. Winning the international tournament ____________ her **reputation** as a world-class athlete.
4. Adrenaline is one of the best-known **stress** ____________.
5. Diabetes is a very common ____________ **disease**.

**D** Complete the passage using the correct form of the words in **bold** from Exercise A.

| inevitable | strengthen | crisis | release | response |
|---|---|---|---|---|

When a person's stress [1] ____________ remains active for a long time, it is [2] ____________ that their health suffers. A 2017 study in Canada suggests that public safety workers, such as firefighters, police officers, and paramedics, are more likely to suffer from mental health disorders compared to the general public. One reason for this could be the potentially traumatic [3] ____________ that these workers face. In recent years, organizations and programs have been created to provide these first responders with the support they need. First responders also use different strategies to cope with stress. Some are able to [4] ____________ their stress by exercising or by talking to trained professionals. Others [5] ____________ their mental resilience by making sure to take time for themselves, to meditate, and to reach out to their peers.

## COMMUNICATE

**E** Work with a partner. Discuss the questions below.

1. Which of the stress-reduction techniques from the infographic have you tried? Which are most effective?
2. Why do you think first responders tend to suffer from chronic stress?

# Viewing and Note-taking

LEARNING OBJECTIVES

- Watch a lecture on the effects of stress
- Use symbols for note-taking
- Listen to identify cause and effect

## BEFORE VIEWING

**A** Look at the symbols in the box below. Which ones do you use the most often? What other symbols do you use? Discuss with a partner.

| | | | |
|---|---|---|---|
| → | leads to, causes | ↓ | decrease |
| ← | results from | ↓↓ | a sharp decrease |
| ≠ | not equal to | # | number |
| ≈ | about, around | / | per, divided by |
| ↑ | increase | Δ | change |
| ↑↑ | a sharp increase | + | and |

**Note-taking Skill**

**Using Symbols**

To show relationships between ideas, it can be easier and quicker to use symbols when you take notes.

**B** Listen to the sentences. Use the symbols in Exercise A to take notes. Then compare your notes with a partner.

E.g. *Studies show that getting good sleep helps reduce stress significantly.*

*Getting good sleep → ↓↓ stress*

**1.** ______________________________

**2.** ______________________________

**3.** ______________________________

## WHILE VIEWING

**C** **LISTEN FOR MAIN IDEAS** Watch Segment 1 of the lecture and circle the correct option to complete the sentences.

**1.** Stress is

**a.** almost always harmful to our health.
**b.** harmful when it lasts for a short time.
**c.** important for our survival in a crisis.

**2.** Acute stress

**a.** is dangerous over the short term.
**b.** does not last for a long time.
**c.** is often harmful to our health.

**3.** Chronic stress

**a.** can last for a short or long period of time.
**b.** can make us sick as it lasts for a long time.
**c.** does not last for long but is very harmful.

**D** ▶ **LISTEN FOR DETAILS** Watch Segment 1 of the lecture again and complete the flow charts to show the cause-and-effect relationships.

**ACUTE STRESS**

| | | | | | | |
|---|---|---|---|---|---|---|
| Brain releases stress [1] ________ = immediate [2] ________ | → | [3] ________ beats faster<br>Blood pressure + breathing rate ↑<br>= more oxygen to [4] ________ + muscles | → | Muscles tense up →<br>run [5] ________ and escape | → | **SAFE** |

**CHRONIC STRESS**

| | | | | | | |
|---|---|---|---|---|---|---|
| Brain keeps [6] ________ stress hormones | → | Blood pressure + heart rate stay [7] ________ | → | Muscles don't relax | → | **NOT SAFE** |

**E** ▶ **IDENTIFY CAUSE AND EFFECT** Watch Segment 2 of the lecture and complete the cause-and-effect relationships below.

1. Nowadays, ________________ → stress
2. Fear of losing a job can last a long time
   → ________________
3. The pressure on the heart over time
   → increased risk of ________________
4. Chronic stress may
   → physical problems including high cholesterol and ________________

**Listening Skill**

**Listening for Cause and Effect**

In a lecture it may be important to understand cause-and-effect relationships. You may need to infer the relationship from the context; other times the speaker will use signal words and phrases.

## AFTER VIEWING

**F** **APPLY** The lecturer states that psychological fears are more likely to cause chronic stress than actual physical danger. Are the fears below based on physical danger (write **D**), or psychological fear (write **P**)? Could any be both? Discuss with a partner.

______ **1.** getting bad grades

______ **2.** getting into a car accident

______ **3.** falling from a great height

______ **4.** giving a speech

______ **5.** losing your cell phone

______ **6.** the dark

______ **7.** spiders

______ **8.** being robbed

# Noticing Language

LEARNING OBJECTIVES

- Notice language for talking about cause and effect
- Explain the causes and effects of two types of stress

## LISTEN FOR LANGUAGE *Describe cause and effect*

**A** Listen to excerpts from the lecture in Lesson B. What phrases does the speaker use to talk about cause and effect? Complete the sentences below.

1. Our stress tends to be __________ psychological fear or worry about the future ...
2. The __________ chronic stress __________ the heart are the easiest to explain ...
3. Chronic stress can __________ the brain, the stomach, and other muscles.

**B** Work with a partner. Look at the signal words and phrases below. What similarities and differences do you notice between them?

| Verb | Noun |
|---|---|
| **affect**<br>*Chronic stress **affects** the body in many ways.* | **effect (of) ... (on) ...**<br>*An **effect of** chronic stress **on** the body is an increased risk of heart disease.* |
| **cause**<br>*Chronic stress can **cause** headaches.* | **cause (of)**<br>*Chronic stress can be the **cause of** headaches.* |
| **result in**<br>*A stress response that lasts too long **results in** chronic stress.* | **result (of)**<br>*Chronic stress is the **result of** a stress response that lasts too long.* |

**Communication Skill**

**Describing Cause and Effect**

Certain verbs and nouns can be used to indicate cause and effect. Pay attention to word order and prepositions (if any).

**C** Rewrite the questions by changing the words in **bold** to their noun or verb forms. Then, with a partner, take turns asking and answering the questions with ideas from the lecture and your own ideas.

1. What are the **causes** of chronic stress?
   What __________?
2. What does too much stress **result in**?
   What are the __________?
3. How does stress **affect** your everyday life?
   What are the __________?

A university student expresses her feelings to a counselor and group members at a therapy session in Texas, U.S.A.

**D** Listen to part of a podcast about how we affect other people's stress and complete the chart below with the effects of each cause.

| Cause | Effect |
| --- | --- |
| smile or kind word | **1.** ______ |
| strong, caring relationships | **2.** ______ |
| frequent unpleasant words | **3.** ______ |
| constant state of stress | **4.** ______ |

## COMMUNICATE

**E** Take notes to prepare for a short oral summary of what happens to your body during acute and chronic stress. Use the cause and effect language from Exercise B.

**F** Work with a partner. Take turns to explain your summaries using your notes from Exercise E. Compare your ideas with the flowcharts in Lesson B. Did you forget anything?

# Communicating Ideas

LEARNING OBJECTIVES

- Use appropriate language for making recommendations
- Collaborate to provide solutions

**ASSIGNMENT**

**Task:** You are going to collaborate with a partner to provide solutions for how two students might reduce their stress levels.

## LISTEN FOR INFORMATION

**A** **LISTEN FOR MAIN IDEAS** Listen to a conversation between Sarah and Wei Ming, two college students. What issues are they discussing? Check (✓) the ones you hear.

**1.** ☐ exams
**2.** ☐ money
**3.** ☐ parties
**4.** ☐ jobs
**5.** ☐ assignments
**6.** ☐ relationships

**B** **LISTEN FOR DETAILS** Listen and complete the flow charts for Wei Ming and Sarah. Use the phrases in the box to help you.

| | |
|---|---|
| **can't focus in class** | **not getting enough sleep** |
| **falling further and further behind** | **takes a job** |
| **no time to study** | **works till late** |

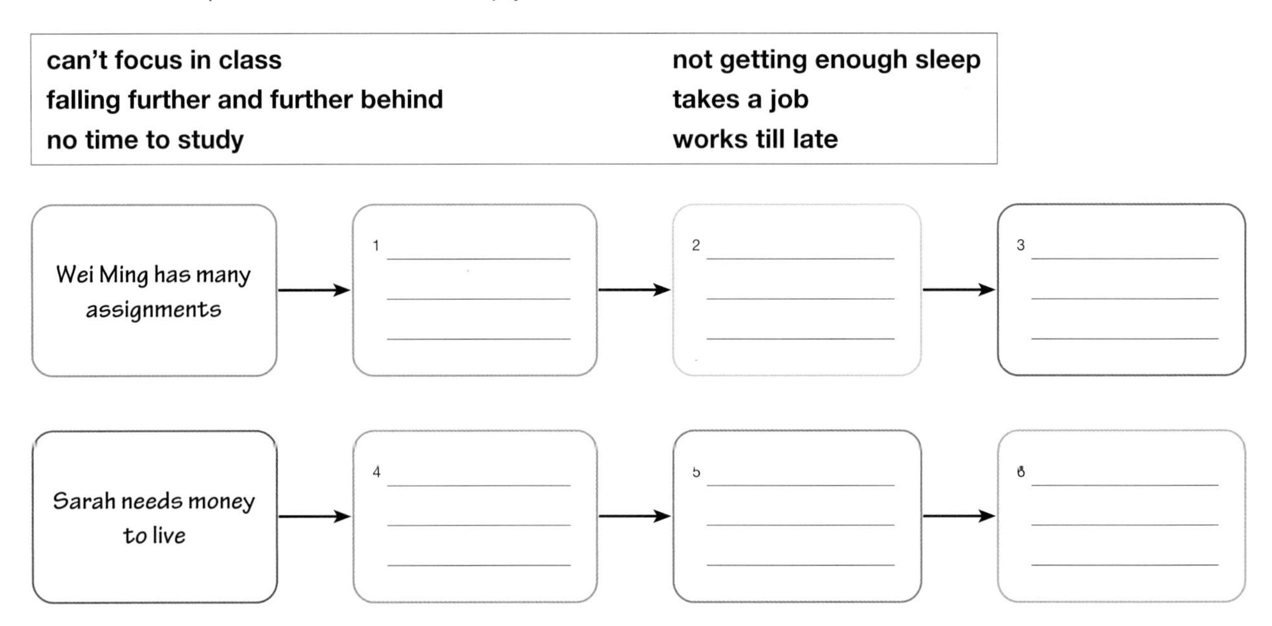


**C** Work with a partner. Compare your notes. Do you think Wei Ming and Sarah are making the right choices?

A student prepares for the first subject of the Annual College Entrance Exams, Harbin, China.

### COLLABORATE

**D** Work with a different partner. What could Sarah and Wei Ming do to deal with their issues and reduce their stress? Suggest some practical solutions.

> I think Sarah could quit her job. She needs the time to study.

> That's true, but what about her everyday costs?

**E** Share your ideas with the class. Explain your reasoning.

## Checkpoint

Reflect on what you have learned. Check your progress.

**I can ...** understand and use words related to stress and anxiety.

| | | | | |
|---|---|---|---|---|
| **chronic** | **crisis** | **enhance** | **hormone** | **inevitable** |
| **mechanism** | **response** | **release** | **reveal** | **strengthen** |

use collocations with words related to stress.

watch and understand a lecture on the effects of stress.

understand and use symbols for note-taking.

listen to identify cause and effect.

notice language for talking about cause and effect.

explain the causes and effects of two types of stress.

communicate and collaborate effectively to provide solutions for stress-related issues.

Yoga is an increasingly popular way to reduce stress and enhance health. Here, more than 2,000 people attend a yoga class at Red Rocks Amphitheatre in Colorado, U.S.A.

# Building Vocabulary

LEARNING OBJECTIVES

- Use ten words related to managing stress
- Use different forms of *confession, heal, compassionate, appreciation,* and *empathy*

## LEARN KEY WORDS

**A** Listen to and read the passage below. Are the tips to reduce stress surprising to you? Discuss with a partner.

**Strategies for Managing Stress**

Whether we're at college, at work, or at home, most of us experience at least a **moderate** amount of stress regularly—but by using some simple strategies, we can increase our **resilience** to stress. One way is to **rethink** the way we view stressful situations. For example, after a bad day, instead of thinking about the experiences that made us feel anxious or unhappy, we should focus on what we learned from them. That way, we can build up the **courage** to face the next day's challenges even if our circumstances remain the same.

Another simple way to reduce stress is through massage. A recent study shows that even a short massage can help to reduce mental as well as physical stress. In the study, **participants** were given a ten-minute massage, and their heart rates were monitored. After the massage, all the participants reported that they felt more relaxed, and their results indicated that they had experienced a physiological reduction in stress.

**B** Work with a partner. Discuss the questions below.

1. Think of a recent stressful experience. Is there something positive you could take from that experience?
2. Aside from massage, what are some simple methods you use to feel more relaxed?
3. Look at the photo. Have you ever tried yoga? Would you like to go to an event like this?

**C** Complete the sentences with the correct form of the words in **bold** from Exercise A.

1. The ____________ in the study showed reduced levels of stress after undergoing the test.
2. She pointed out that it is important to ____________ our attitude toward stress.
3. Sometimes I don't have the ____________ to speak up during a meeting.
4. There are various techniques that you can use to improve your ____________ to stress.
5. The students report experiencing a ____________ increase in stress since the start of the school year.

**D** Read the excerpts from Kelly McGonigal's TED Talk in Lesson F. Then choose the options that are closest to the meaning of the words in **bold**.

1. But first, I want you to make a little confession to me. In the past year, I want you to just raise your hand if you've experienced relatively little stress.

   When you make a **confession**, you ______.

   **a.** admit something **b.** tell a story **c.** lie about something

2. ... oxytocin helps heart cells regenerate and heal from any stress-induced damage.

   When you **heal,** you ______.

   **a.** become healthy again **b.** become clean again **c.** become bigger

3. The compassionate heart ... finds joy and meaning in connecting with others.

   **Compassionate** people ______.

   **a.** are very sociable **b.** show sympathy for others **c.** think deeply

4. ... this science has given me a whole new appreciation for stress.

   **Appreciation** means ______.

   **a.** enjoyment **b.** thankfulness **c.** curiosity

5. Oxytocin makes you crave physical contact with your friends and family. It enhances your empathy. It even makes you more willing to help and support the people you care about.

   **Empathy** means ______.

   **a.** ability to remember facts **b.** ability to understand other people's feelings **c.** ability to feel sorry for others

**E** Complete the chart with the correct form of the words. Use a dictionary if necessary.

| Noun | Verb | Adjective |
|---|---|---|
| confession | | X |
| | heal | |
| | X | compassionate |
| appreciation | | |
| empathy | | |

## COMMUNICATE

**F** Work with a partner. Discuss the questions below. Use the words in **bold** in your answers.

1. What jobs require **empathy**? Why?
2. Do you know anyone who has a lot of **courage**? What courageous things have they done?
3. Do you consider yourself **resilient** to stress? Why, or why not?

# Viewing and Note-taking

LEARNING OBJECTIVES

- Watch and understand a talk about our attitudes toward stress
- Notice the use of thought groups



## TEDTALKS

Stanford University psychologist **Kelly McGonigal** is the author of a book titled *The Upside of Stress: Why Stress Is Good for You, and How to Get Good at It.* In her TED Talk, *How to Make Stress Your Friend*, she encourages us to rethink our attitudes toward stress.

### BEFORE VIEWING

**A** Read the information about Kelly McGonigal. In what ways might stress be good for us? Discuss your ideas with a partner.

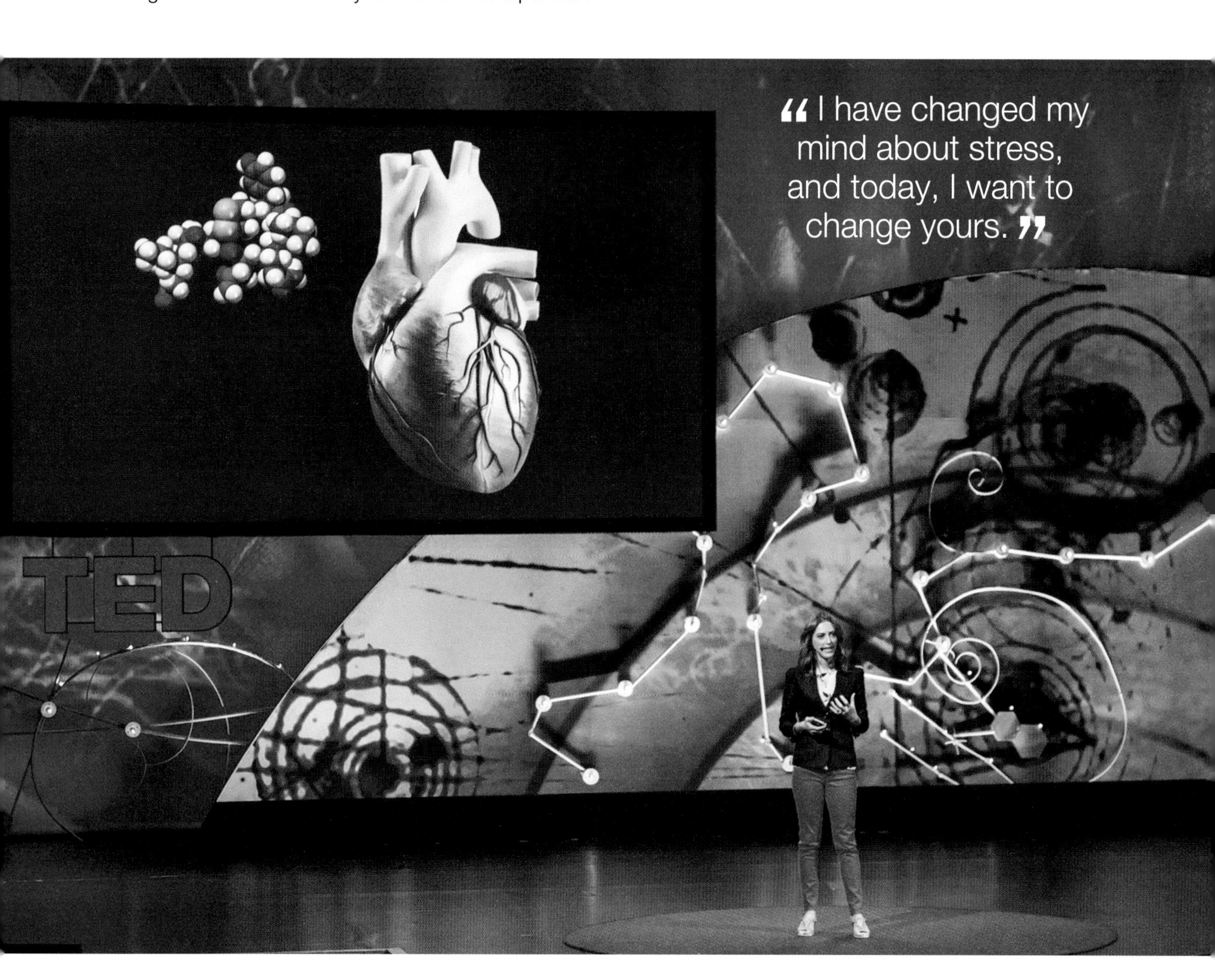


"I have changed my mind about stress, and today, I want to change yours."

## WHILE VIEWING

**B** ▶ **LISTEN FOR MAIN IDEAS** Watch Segment 1 of Kelly McGonigal's TED Talk. Circle the best answer to each question.

**1.** What does McGonigal think she has done wrong?

**a.** She has been teaching people that stress is good for them, but it is not.

**b.** She has put people's health in danger by telling them that stress is bad for their health.

**c.** She has continued working in a job that causes her a lot of stress.

**2.** What did McGonigal learn that changed her mind about stress?

**a.** Stress has both harmful and beneficial effects on the body.

**b.** Stress only affects the people who think a lot about it.

**c.** Stress can be harmful depending on the way people think about it.

**C** ▶ **LISTEN FOR DETAILS** Watch Segment 2 of McGonigal's TED Talk and complete the flow chart.

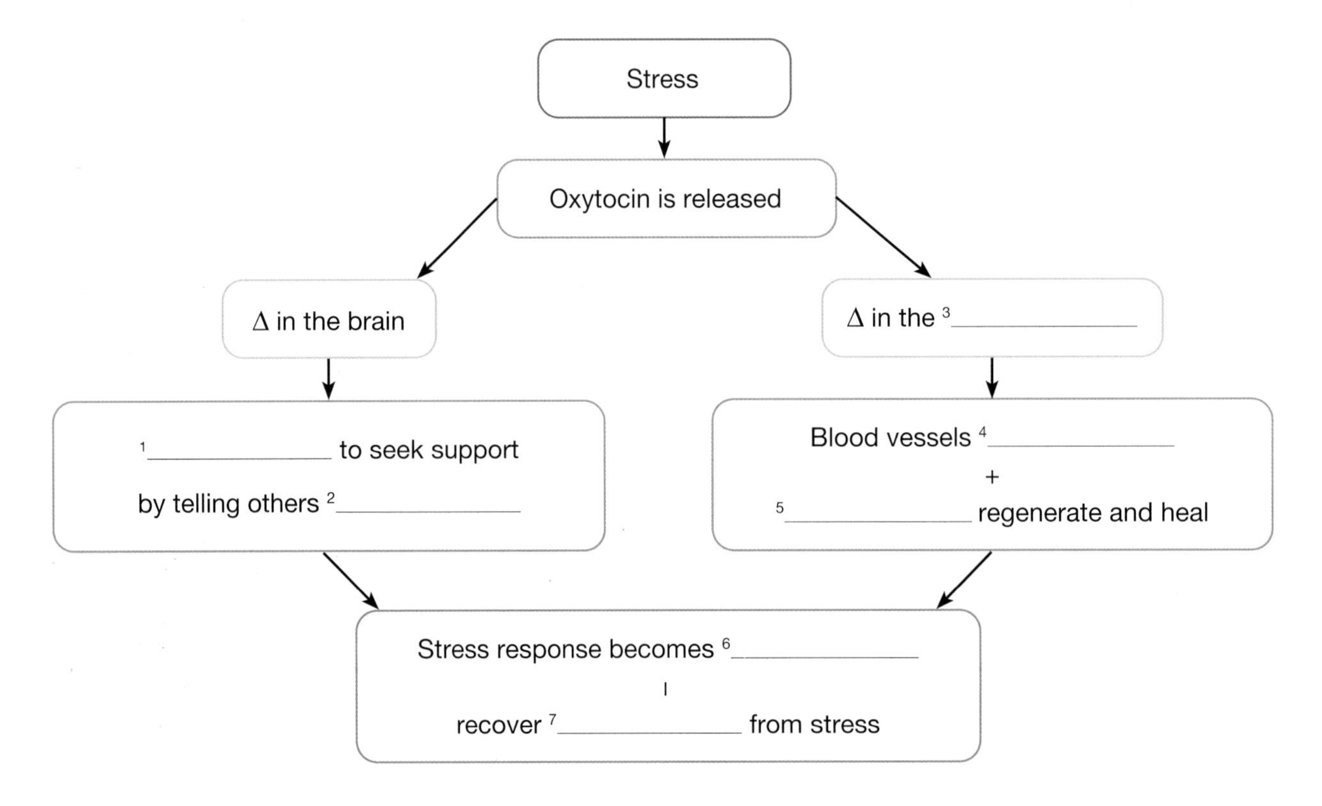


**WORDS IN THE TALK**

*cardiovascular* (adj) related to the heart or blood vessels

*cell* (n) the smallest part of an animal or plant's body

## AFTER VIEWING

**D** **SUMMARIZE** Look at the chart below and summarize the two studies in McGonigal's TED Talk by using the phrases in the word box.

| | | |
|---|---|---|
| **helping out** | **risk of dying** | **make us more resilient** |
| **think about** | **experienced in the last year** | **caring for others** |

| | Study 1 | Study 2 |
|---|---|---|
| **Questions asked** | How much stress have you [1] ______________? | |
| | Do you believe stress is harmful for your health? | How much time have you spent [2] ______________ friends, neighbors, or people in your community? |
| **Findings** | People who experienced a lot of stress and who believed that stress was bad for their health increased their [3] ______________ by 43%. | People who experienced a lot of stress but who spent time [4] ______________ did not have a stress-related increase in dying. |
| **Conclusion** | The way we [5] ______________ stress matters. | Social contact and social support help to [6] ______________. |

## PRONUNCIATION *Thought groups*

**E** Read and listen to the example of thought groups from the lecture in Lesson B.

*You experience acute stress in certain situations / usually when something unexpected happens / causing you to feel threatened.*

**F** Read the sentences below. Put a slash ( / ) at the end of each thought group. Then, with a partner, take turns reading the sentences according to your markings.

> **Pronunciation Skill**
> **Using Thought Groups**
> English speakers group words into segments of meaning called thought groups. Thought groups help the listener understand and process information. There is a small break or pause at the end of each thought group. Long sentences might contain several thought groups. Short sentences might contain just one or two.

1. Talking with others about things that stress you out can help you calm down and relieve your anxiety.
2. Keeping a stress journal about what causes you stress and how you feel and react to it is one way of identifying the stressors in your life.

UNIT 4

E F G H

# Thinking Critically

LEARNING OBJECTIVES

- Interpret an infographic about the different causes of stress
- Synthesize and evaluate ideas about stress levels

## ANALYZE INFORMATION

**A** Look at the infographic below. Which of the questions (1–4) does it answer? Check (✓) two questions.

**1.** ☐ What are some major causes of stress?

**2.** ☐ What can people do to reduce their stress overall?

**3.** ☐ Why do family responsibilities cause stress?

**4.** ☐ What are some reasons for work-related stress?

# What causes stress?

Commonly cited as one of the top stressors around the world, work plays a large role in determining our physical and mental well-being. Greater workloads and longer working hours have resulted in higher numbers of people reporting feelings of stress.

A 2017 estimate by the American Institute of Stress indicated that job stress alone costs U.S. industry more than $300 billion a year.

**MONEY / WORK STRESS**

- Bills
- Job
- Job review
- Employer
- Workload
- Deadlines
- Mondays
- Duties
- Promotion
- Retirement

Source: American Psychological Association

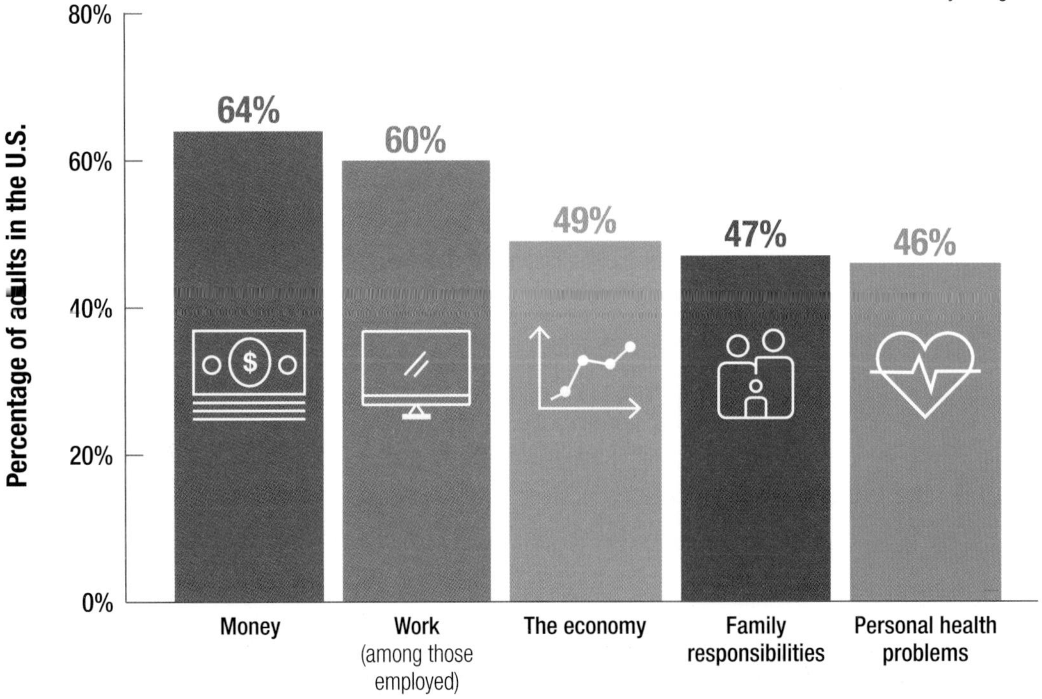

Percentage of adults in the U.S.

80%
60%
40%
20%
0%

| Money | Work (among those employed) | The economy | Family responsibilities | Personal health problems |
|---|---|---|---|---|
| 64% | 60% | 49% | 47% | 46% |

**Top causes of stress in 2021**

**B** **LISTEN FOR DETAILS** Listen to part of a lecture about stressors and mental health. Check (✓) the topics the speaker talks about.

- ☐ financial pressures
- ☐ work problems
- ☐ getting married
- ☐ starting a new job
- ☐ moving to a new home
- ☐ chronic pain/illness
- ☐ caring for an ill loved one
- ☐ family problems
- ☐ world issues
- ☐ exams

**C** Discuss the questions below with a partner.

1. How can work stress cost businesses money? Think of a few different ways.
2. What do you think was the main message of the speaker in Exercise B?
3. Share an experience you have had with one of the types of stress mentioned in Exercise A or B. How did (or do) you deal with it?

## COMMUNICATE *Synthesize and evaluate ideas*

**D** Think about what you have learned about stress in this unit. Which scenario below would you find the most stressful? Rank the scenarios from 1 (the most stressful) to 3 (the least stressful).

_______ **a.** You work as an administrative assistant. You work 50 hours a week for a large company. You also study for a diploma on weekends. Twice a month, you volunteer at a non-profit organization.

_______ **b.** You work as the manager of a small travel agency. You work in the office for 40 hours a week for three weeks a month. You spend the fourth week on business trips organized by your boss to give presentations and meet new customers.

_______ **c.** You work as the operations manager of a large software company. You work 60 hours per week, set your own schedule, and often work from home. Your family will be moving to a new home in two months' time.

**E** Work with a partner. Compare your answers. Take turns to explain your rankings and discuss your ideas.

I think scenario A is the most stressful. I don't think I'd be able to work and study at the same time.

I'm not sure. I think B is probably the most stressful. The business travel would be tiring, and ...

UNIT 4

E F G H

# Putting It Together

LEARNING OBJECTIVES

- Research and plan a survey on stress, and present on the findings
- Use a varied pace to emphasize or create suspense in a presentation

**ASSIGNMENT**

**Group presentation:** Your group is going to conduct a survey on stress and give a presentation on the results.

## PREPARE

**A** Work with your group to prepare for your presentation. Follow these steps.

1. Look at the survey below. Your group should interview at least eight people. Ask the questions orally. After each question, ask at least one follow-up question to find out more. Ask permission to record your interviews and/or take notes.
2. Analyze the results. What conclusions, if any, can you make? (e.g., "Most of the people who reach out to other people for support turn to a family member.")
3. Organize your presentation. In it, you should:
   - Report what the results were (e.g., "6 out of 8 people answered 'Agree' to Question 1").
   - Draw logical conclusions based on the results.
   - Discuss how the results support (or don't support) what you've learned in this unit.
4. Assign a part of the presentation to each member of the group.

**Stress Survey**

I am conducting a survey on stress for a class presentation. I would appreciate it if you could help me by answering the following questions. Thank you.

Please respond to the statements below by using the following:

**SA** (Strongly agree) **A** (Agree) **N** (Neutral) **D** (Disagree) **SD** (Strongly disagree)

______ **1.** I feel stressed every day.

______ **2.** When I am stressed, I reach out to other people for support.

______ **3.** I know what the top causes of stress in my life are.

______ **4.** I know of more than one way to reduce my stress.

______ **5.** I usually reduce my stress in a healthy way.

**B** Look back at the vocabulary, pronunciation, and communication skills you've learned in this unit. What can you use in your part of the presentation? Note any useful language below.

---

---

**C** Below are some ways to vary your pace. Think about how you can use these ideas in your presentation to make it more engaging.

- Pause a little longer than usual before and after the information you want to emphasize.
- Speed up a little on information that is not so important.
- Slow your pace just before the information you want to emphasize, then speak slower when you give that information.

**D** Practice your presentation. Make use of the presentation skill that you've learned.

**Presentation Skill**

**Varying Your Pace**

In Kelly McGonigal's TED Talk, she varies her pace at different points. Varying the pace of your speech in a presentation can help you emphasize an important, interesting, or unexpected piece of information. It can also help you create drama and suspense so that your audience is eagerly waiting for what you will say next.

## PRESENT

**E** Give your presentation to another group. Watch their presentation and evaluate them using the Presentation Scoring Rubrics at the back of the book.

**F** Discuss your evaluation with the other group. Give feedback on two things they did well and two areas for improvement.

# Checkpoint

Reflect on what you have learned. Check your progress.

**I can ...** ☐ understand and use words related to managing stress:

| | | | | |
|---|---|---|---|---|
| **appreciation** | **compassionate** | **confession** | **courage** | **empathy** |
| **heal** | **moderate** | **participant** | **resilience** | **rethink** |

☐ use different forms of *confession, heal, compassionate, appreciation,* and *empathy.*

☐ watch and understand a talk about our attitudes toward stress.

☐ notice and use thought groups.

☐ interpret an infographic about the different causes of stress.

☐ synthesize and evaluate ideas about stress levels.

☐ use a varied pace to emphasize or to create suspense in a presentation.

☐ research and plan a survey on stress, and present on the findings.

Local volunteers build a new school in a village in Yinchuan, Ningxia Hui, China.

# 5

# A Helping Hand

## Q When does helping really help?

In the photo, a group of volunteers work together to build a new school in a village in China. It's hard to see how a project like this, funded by a French NGO, could have downsides. However, helping people isn't always straightforward, and even the best intentions can sometimes backfire. It's therefore important not just to want to help people, but to consider how best to do it. In this unit we explore various ways of helping others, and the unexpected effects "helping" can have.

## THINK and DISCUSS

1 Look at the photo and read the caption. Why do you think the volunteers are helping to build the school? What do they get out of it?

2 Look at the essential question and the unit introduction. How might helping someone do more harm than good?

UNIT 5

A B C D

# Building Vocabulary

LEARNING OBJECTIVES

- Use ten words related to charitable giving
- Use the prefix *under-*

## LEARN KEY WORDS

**A** Listen to and read the information below. Discuss with a partner.

1. What does the WGI measure?
2. How are WGI rankings determined?

# The Worldwide Giving Index

A volunteer at a food bank in Winford, U.K.

Every year, the Charities Aid Foundation publishes the World Giving Index (WGI), a report based on the largest survey of charitable giving worldwide. The report is not based on the money a country **donates** or people's **motives** for giving. Rather, it looks at the percentage of people in each country who claim to have donated **funds**, volunteered time, or helped a stranger in the previous month.

One might **assume** that the most **generous** nations were the wealthiest and therefore most **capable** ones, but that would be incorrect. For many years, the top ten ranking has included countries at all stages of economic development. In fact, in 2020, the top five spots were taken mostly by less wealthy countries. Neil Heslop, who **took over** as CEO of the Charities Aid Foundation in 2020, believes that this was down to the COVID-19 pandemic.

According to Heslop, charitable activity in wealthy countries relies heavily on an **infrastructure** of fundraisers, thrift shops, community **allies**, and volunteers. However, the effectiveness of this infrastructure was **undermined** during the pandemic when many volunteers and charity workers were no longer able to participate as effectively.

## Most Charitable Countries

### 2009–2018

Top 5 countries (aggregated data from 2009–2018)

| | 1 United States | 2 Myanmar | 3 New Zealand | 4 Australia | 5 Ireland |
|---|---|---|---|---|---|
| H | 72% | 49% | 64% | 64% | 62% |
| C | 61% | 81% | 65% | 68% | 69% |
| V | 42% | 43% | 41% | 37% | 38% |
| AVG | **58%** | **58%** | **57%** | **56%** | **56%** |

### 2020

Top 5 countries in 2020

| | 1 Indonesia | 2 Kenya | 3 Nigeria | 4 Myanmar | 5 Australia |
|---|---|---|---|---|---|
| H | 65% | 76% | 82% | 51% | 57% |
| C | 83% | 49% | 33% | 71% | 61% |
| V | 60% | 49% | 42% | 31% | 30% |
| AVG | **69%** | **58%** | **52%** | **51%** | **49%** |

**H** Helped a stranger **C** Donated money to a charity **V** Volunteered time to an organization

Source: Charities Aid Foundation Index, 2021

**B** Match the correct form of each word in **bold** in Exercise A with its meaning.

1. ______ likely to give more than is expected
2. ______ to make something weaker or less strong
3. ______ to accept that something is true without proof
4. ______ having the ability to do something well
5. ______ to gain control of a situation from someone else
6. ______ the facilities an organization or system needs to run
7. ______ to give money or goods, for example to a charity
8. ______ someone who works with you, not against you
9. ______ reason for doing something
10. ______ money that has been set aside for a specific purpose

**C** The word *under* is sometimes used as a prefix that means *beneath* or *less than*. Complete the sentences with the correct words from the box below.

| **underestimate** | **underprivileged** | **undermine** | **underground** |
|---|---|---|---|

1. He tried to ______ my authority by complaining to my coworkers.
2. ______ children often do not have the opportunity to go to college.
3. Electrical wires are sometimes buried ______ so that they aren't visible.
4. He's quiet, but don't ______ him. He's good at what he does.

**D** Complete the passage using the correct form of the words in **bold** from Exercise A.

Why do people [1] ______ to charities? You might [2] ______ it's because they care, and are [3] ______ in nature. But some people give because they want something back, like media attention or a tax break. Should people's [4] ______ really matter? Charities need [5] ______ to operate. Should they care about why their [6] ______ give, so long as they're giving?

## COMMUNICATE

**E** Work with a partner. Discuss the questions below.

1. Think about how the WGI measures which countries are the most charitable. What are the pros and cons of this method?
2. Are people's motives for donating to charities important? Why, or why not?

UNIT 5

# Viewing and Note-taking

LEARNING OBJECTIVES

- Watch a video podcast about what people do to help others
- Note down causes and their effects
- Recognize signpost questions

## BEFORE VIEWING

**A** Read the sentences below. Underline the causes and effects. Does each signal word or phrase introduce the cause or the effect?

**Words and phrases for cause and effect**

| | |
|---|---|
| due to | Due to the earthquake yesterday, the event was canceled. |
| so | There was an earthquake yesterday, so the event was canceled. |
| as a result | There was an earthquake yesterday. As a result, the event was canceled. |
| result in | Yesterday's earthquake resulted in the cancellation of the event. |
| because | The event was canceled because there was an earthquake yesterday. |

**Note-taking Skill**

**Noting Down Cause and Effect**

It's sometimes useful to identify and label causes and their effects in your notes. Often speakers will use signal words and phrases to indicate cause and effect, but sometimes you will need to infer the relationship.

**B** **LISTEN FOR DETAILS** You are going to listen to a description of GiveDirectly, a charity started by aid worker Joy Sun. Complete the sentences below and discuss them with a partner.

| Causes | | Effects |
|---|---|---|
| 1. ______ is given to people directly. | → | People can make their own ______ based on their own ______. |
| 2. ______ can choose to buy higher quality food. | → | Their ______ are born ______. |
| 3. Families can start ______ with the money they get. | → | They are able to generate ______ and meet local ______. |

## WHILE VIEWING

**C** **PREDICT** Watch part of a video podcast about what people do to help others. What sort of help do you think the speaker is going to talk about? Share your ideas with a partner.

Donated goods need to be transported to disaster sites and containers like these need to be unloaded.

**D** ▶ **LISTEN FOR MAIN IDEAS** Watch Segment 1 of the video podcast. What problem does it describe?

1. Most people respond too slowly to disasters.
2. Only a small percentage of people respond to disasters.
3. People often respond to disasters in ways that don't help.

**E** ▶ **LISTEN FOR CAUSE AND EFFECT** Watch Segment 1 again. Note down the effects of each cause.

| Causes | Effects |
|---|---|
| People see images of loss and suffering after a disaster. | |
| People donate too many things after a disaster. | |

**F** ▶ **LISTEN FOR SIGNPOST QUESTIONS** Watch Segment 2 of the video podcast. Write notes to answer the four questions that were asked in the video.

| | |
|---|---|
| **1.** But just what are these problematic goods? | **2.** How do such large, unusual donations come about? |
| **3.** And what exactly is this right thing? | **4.** But what about those businesses that continue to behave badly? |

**Listening Skill**

**Recognizing Signpost Questions**

Speakers often introduce new ideas–like causes and effects–by asking questions. Listening for signpost questions like *So how did ...?* or *And what about ...?* can help you follow a speaker's message.

## AFTER VIEWING

**G** **APPLY** Work with a partner. Discuss the questions below.

1. Do you agree with the podcaster's point of view? Why, or why not?
2. How do you think Joy Sun, the founder of GiveDirectly, would react to the podcaster's point of view?
3. When might it be better to donate goods rather than money? Why?

# Noticing Language

LEARNING OBJECTIVES

- Notice the use of emphatic stress to appeal to emotions
- Appeal to people's emotions

## LISTEN FOR LANGUAGE *Appeal to emotions*

**A** Listen to excerpts from the video podcast in Lesson B. Which words get emphatic stress? Underline them.

1. But how much good does our help really do?
2. In fact, they often undermine humanitarian efforts, due to the time and effort needed to deal with all the donations.
3. The problem is all too familiar to disaster specialists, who see it happen over and over again.

**Communication Skill**

**Appealing to Emotions**

Speakers often use emphatic stress when they are appealing to people's emotions, for example, when they want to call attention to something, or convince someone to do–or not do–something. Emphatic stress involves lengthening the vowel in the stressed syllable and changing the pitch on the stressed word.

**B** Practice reading the sentences from Exercise A aloud. What is the speaker's intention for each sentence?

The speaker uses emphatic stress to ...

1. **a.** urge people to volunteer and donate more.
   **b.** get listeners to reconsider a common opinion.
2. **a.** stress that humanitarian work requires time and effort.
   **b.** point out that donations can do more harm than good.
3. **a.** express frustration that the same thing keeps happening.
   **b.** emphasize how difficult disaster specialists' jobs are.

**C** Read what the speaker says in each situation below. Circle the word you think the speaker will emphasize. Then listen and check.

1. A fundraiser wants to convince people to make donations. She is making a personal, direct appeal to every person who is watching or listening.

   "We need you to help us in our fundraising efforts."
2. A lecturer wants her students to pay close attention to important information. She does this by emphasizing that the information will definitely be on the test.

   "This will be on the test."
3. A girl is trying to convince her friend to go skydiving with her. He has no intention of doing it, under any circumstances.

   "There's no way I'm doing that!"
4. A pilot gets sick and needs his less experienced copilot to fly the plane. It is very important that the copilot follow his instructions.

   "Do exactly what I tell you, OK?"

Malala Yousafzai delivers an impassioned speech at a Youth Assembly event at the U.N. headquarters in New York, U.S.A.

**D** Work with a partner. Read each scenario and write a one-sentence emotional appeal. Underline the word(s) in the sentence that should receive emphatic stress.

1. There has been a natural disaster and the victims need help immediately. You are an aid worker who will be addressing the public. Make an emotional appeal for the public's help.

   Your emotional appeal: ______________________________

2. Your friend is struggling with an assignment that's due soon, but he doesn't want anyone to help him. Make an emotional appeal to convince him to accept your help.

   Your emotional appeal: ______________________________

3. You're going to college, but you want to major in something your parents don't approve of. Make an emotional appeal to convince them to let you follow your passion.

   Your emotional appeal: ______________________________

**E** Read your sentences aloud. Discuss whether you think your emotional appeals are effective and whether you chose the correct word to receive emphatic stress.

## COMMUNICATE

**F** Prepare an emotional appeal. Choose **one** of the statements and write a few sentences about why we should try to help people this way. Underline words that should receive emphatic stress.

**We should help people in need by:**

- donating money to a charitable organization.
- giving money directly to individuals.
- volunteering time and providing direct assistance.
- reaching out to others and getting more people to help.

**G** Work with a partner. Take turns making your emotional appeals from Exercise F. Did your partner make a good argument? Did they use emphatic stress?

# Communicating Ideas

LEARNING OBJECTIVES

- Use appropriate language for supporting your viewpoint
- Collaborate to make an argument by appealing to emotions

**ASSIGNMENT**

**Task:** You are going to collaborate with a partner to support a viewpoint you feel strongly about and use your ideas to convince others.

## LISTEN FOR INFORMATION

**A** **LISTEN FOR MAIN IDEAS** Listen to a student talking about a cause she believes in. What is the purpose of her talk?

1. To get people to visit the bank's website and learn more about it
2. To convince listeners that microcredit is effective and sustainable
3. To encourage people to support a bank that helps people escape poverty

**B** **LISTEN FOR DETAILS** Listen again. Complete the notes below.

1. Microcredit refers to ______________________________

______________________________

2. Microcredit is effective because ______________________________

______________________________

3. Microcredit is sustainable because ______________________________

______________________________

**C** Read the pairs of sentences about Grameen bank. Which is the cause and which is the effect? Write **C** or **E**.

______ **1.** Poor people have the means to escape poverty.
______ Grameen provides very small loans to people.

______ **2.** Grameen loans money to people in groups of five.
______ People repay their loans because of peer pressure.

______ **3.** Two people in a group of five make payments on time.
______ The remaining three people are allowed to get loans.

______ **4.** It can continue to help people to escape poverty.
______ Grameen's lending method is sustainable.

## COLLABORATE

**D** Work with a partner. Read the statements in the box. Choose **one** that you strongly agree or disagree with. List reasons to support your position.

- You should help someone you know before you help a stranger.
- Helping animals is just as important as helping humans.
- Everyone should donate at least 2% of their income to charity.
- People who help others are happier than those who don't.

Reason 1: ______________________________

Reason 2: ______________________________

Reason 3: ______________________________

**E** Work with a partner. Write two sentences supporting your position that you could use to make an emotional appeal. Underline the words you think should receive emphatic stress.

Sentence 1: ______________________________

Sentence 2: ______________________________

**F** Work with a new partner. Convince them to agree with your position. Try to appeal to their emotions.

# Checkpoint

Reflect on what you have learned. Check your progress.

**I can ...** understand and use words related to charitable giving.

| | | | | |
|---|---|---|---|---|
| **ally** | **assume** | **capable** | **donate** | **funds** |
| **generous** | **infrastructure** | **motive** | **take over** | **undermine** |

use the prefix *under-*.

watch and understand a video podcast about what people do to help others.

make notes about causes and their effects.

listen and recognize signpost questions.

notice and use emphatic stress to appeal to emotions.

use appropriate language for supporting a viewpoint you feel strongly about.

collaborate and communicate effectively to make a convincing argument.

A Peace Corps volunteer teaches English to a class of young boys in Antakya, Turkey.

# Building Vocabulary

LEARNING OBJECTIVES

- Use ten words related to voluntourism
- Understand the antonyms of *external*, *inclusive*, *dependence*, and *outsider*

## LEARN KEY WORDS

**A** Listen to and read the passage below. What is voluntourism? What is the best way to ensure that it does more good than harm?

**Voluntourism: More harm than good?**

Voluntourism—or the performing of volunteer work as a form of tourism—is a topic that often sparks debate. For those volunteering, the experience can be life-changing. However, for local communities, the help received from these short-term volunteers isn't always appreciated.

In many cases, voluntourism has been known to do more harm than good. Sometimes, volunteers with little or no experience in humanitarian or **conservation** work actually slow down progress and bring up costs. And other times, volunteers who do not understand local cultures end up harming the organization's relationship with local communities and **authorities**.

So is voluntourism bad? Not always. When volunteers' abilities match the tasks they're doing, or when the tasks do not require specialized skills, results can be positive. But to ensure this, volunteers need to do their part by learning about the organization they'll be working for and its **expectations**, the work they'll be doing, and the organization's **policies** regarding volunteering.

Organizations should also provide training and **guidance** on culturally appropriate behavior. No matter how prepared volunteers may be, they are still **outsiders** who need support in navigating cultural differences.

**B** Work with a partner. Think about voluntourism. Discuss the questions below.

1. What are some of the motives that people might have for volunteering in foreign countries?
2. Look at the photo. What do you think a volunteer needs to know to be helpful in this context?

**C** Match the correct form of each word in **bold** in Exercise A with its meaning.

1. ____________________ a plan or rule used to make decisions, especially in government or business
2. ____________________ official groups who make rules or laws, and ensure others follow them
3. ____________________ what someone hopes or believes will happen
4. ____________________ people not from a group or community, who might have different values
5. ____________________ help and advice, often from someone you look up to
6. ____________________ protecting the earth's natural resources for future generations

**D** Read the excerpts from Asha de Vos's talk in Lesson F. Choose the options that are closest to the meaning of the words in **bold**.

*"... we need to make marine conservation more* ***inclusive*** *and* ***equitable****. After all, a handful of people from one part of the world can't resolve all the problems of our oceans ..."*

1. Something is **inclusive** if it ______.
   **a.** involves people from different backgrounds **b.** allows only certain people to participate **c.** is cheap and easy to carry out
2. If something is **equitable**, it is ______.
   **a.** well liked **b.** harmful **c.** fair

*"... researchers ... come to countries like mine, do research, and leave without any investment in the local capacity or infrastructure. It creates a* ***dependency*** *on* ***external*** *expertise and makes it unsustainable in the long term."*

3. A **dependency** refers to something that one ______.
   **a.** needs **b.** respects **c.** trusts
4. The word **external** refers to ______.
   **a.** someone who's helpful **b.** something that's outside **c.** something not allowed

**E** The words in the box below are antonyms. Complete the sentences below with the correct words.

| external/internal | inclusive/exclusive | dependence/independence | outsider/insider |
|---|---|---|---|

1. This is a very ______ organization—it's extremely difficult to become a member.
2. I quit because there was a lot of ______ conflict. The leaders of the charity couldn't agree on anything, so nothing got done.
3. We don't have enough knowledge about the local culture. We need advice from an ______.
4. It is essential that human rights organizations maintain their ______ from local governments.

## COMMUNICATE

**F** Note an example next to each prompt below.

1. **Guidance** you received from a teacher ______
2. A **conservation** project ______
3. A local **authority** in your city ______

**G** Work with a partner. Take turns sharing your examples in Exercise F. Respond to your partner's ideas and ask follow-up questions.

My high school history teacher once gave me some useful guidance. She told me to always remember that there are two sides to a story.

Interesting. How did that help you?

# Viewing and Note-taking

LEARNING OBJECTIVES

- Watch and understand a talk about the importance of local representation
- Notice how stress is used within thought groups



**NATIONAL GEOGRAPHIC EXPLORER**

Marine biologist **Asha de Vos** is the founder of the first marine conservation research and education organization in Sri Lanka, Oceanswell. In her talk *A Hero on Every Coastline*, she talks about the problem of "parachute science" and advocates for the greater involvement of local communities in solving global issues.

## BEFORE VIEWING

**A** Read the information about Asha de Vos. "Parachute science" is when scientists come to a developing country, do research and leave, without investing in and involving local people. Why might this be a problem? Discuss with a partner.

## WHILE VIEWING

**B** ▶ **LISTEN FOR MAIN IDEAS** Watch Segment 1 of Asha de Vos's talk. How do you think she felt about the experience and why?

______________________________________________

______________________________________________

**C** ▶ **LISTEN FOR CAUSE AND EFFECT** Watch Segment 1 again. Then complete the chain of cause and effect.

De Vos made a discovery about ______________.

→ De Vos wrote to ______________ for ______________ to kickstart her research.

→ They offered her support, but only if they ______________ the project.

→ De Vos ______________ the offer and ______________ for five years to start her own local project.

→ We now know more about the subject thanks to de Vos's project.

→ De Vos's feelings about ______________ remain with her to this day.

**D** ▶ **LISTEN FOR DETAILS** Watch Segment 2 of de Vos's talk. Then read the questions and check (✓) the correct answers. Each question may have more than one answer.

**1.** What are some common problems with parachute science?

**a.** ☐ It often provides less funding than initially promised.

**b.** ☐ It creates an unequal power balance between locals and outsiders.

**c.** ☐ It makes local researchers dependent on outsiders for help.

**2.** What allies has de Vos found?

**a.** ☐ people who have encountered parachute science and resisted it

**b.** ☐ people who have practiced parachute science in the past and now realize it's wrong

**c.** ☐ a handful of people capable of resolving the problems of the oceans

**3.** Why did de Vos start Oceanswell?

**a.** ☐ to ensure that the heroes who protect the oceans are not forgotten

**b.** ☐ to ensure that the next generation of Sri Lankans will continue to protect the oceans

**c.** ☐ to promote a deeper understanding of our oceans and coastlines

**WORDS IN THE TALK**

*lockdown* (n) a security measure which restricts people from moving freely within or in and out of an area
*poop* (n) an informal term for feces

**E** **LISTEN FOR DETAILS** Watch Segment 3 of the video and circle if the statements are true (**T**) or false (**F**). Discuss your answers with a partner.

1. Small-scale fishers are a small part of the fishing community in Sri Lanka. **T** **F**
2. It was difficult for de Vos to find people to help with her research of small-scale fishers. **T** **F**
3. The participants were probably more honest than they would be with outsiders. **T** **F**
4. Other countries could probably have helped save the stranded pilot whales. **T** **F**
5. The rescue of the whales was a success. **T** **F**

## AFTER VIEWING

**F** **EVALUATE** Work with a partner. Discuss the questions below.

1. Do local researchers in less developed countries need help from foreign organizations? Why, or why not?
2. How should researchers operate when conducting research or conservation projects in foreign countries?

## PRONUNCIATION *Stress inside thought groups*

**G** Listen to and read an excerpt from de Vos's talk. Notice the thought groups and the word she stresses inside each group. Then analyze the sentences below in the same way. Use a slash (/) to separate the thought groups, and underline the stressed word in each group.

*It creates a dependency / on external expertise / and makes it unsustainable / in the long term.*

1. Asha de Vos is a scientist from Sri Lanka whose name is associated with marine biology and conservation.
2. In 2003, shortly after graduating with a degree in marine biology at the age of 24, Asha de Vos made a discovery that would transform her career.

### Pronunciation Skill

**Using Stress Inside Thought Groups**

English speakers often organize sentences into thought groups. Usually, each thought group has one focus word that is stressed more than others. Often, this word is the last content word of the group. Speakers also tend to pause briefly between thought groups.

**H** Work with a partner. Imagine you are a marine biologist speaking to other marine scientists. Make an emotional appeal. Use a slash (/) to separate the thought groups in the sentences below. Underline the words that receive normal stress, and circle words you think should receive emphatic stress. Take turns reading the sentences.

1. It is long past time to put an end to parachute science.
2. As protectors of the oceans, we have a responsibility to support local conservation efforts.
3. Our goal should be to have local heroes on every coastline, fighting for our oceans.

# Thinking Critically

LEARNING OBJECTIVES

- Interpret an infographic about how charitable organizations spend their money
- Synthesize and evaluate ideas about different ways to help

## ANALYZE INFORMATION

**A** Look at the infographic below. What types of overheads do the following examples refer to? Write *rent*, *infrastructure*, *manpower*, or *utilities*. Discuss your ideas with a partner.

**a.** hiring staff and paying workers' salaries ____________

**b.** leasing space for offices, warehousing, etc. ____________

**c.** gas, water, electricity, internet, etc. ____________

**d.** physical equipment, software, vehicles, etc. ____________

**B** For every $100 in donations received, which organization spends the most and least on the following? Circle **RT** (Rainforest Trust), **DB** (Doctors without Borders), or **HK** (Helen Keller International).

| | Most | | | Least | | |
|---|---|---|---|---|---|---|
| **1.** Programs | RT | DB | HK | RT | DB | HK |
| **2.** Fundraising | RT | DB | HK | RT | DB | HK |
| **3.** Overheads | RT | DB | HK | RT | DB | HK |

## Choosing a non-profit organization: How do you decide?

Before deciding which non-profit organization to support, many people find it useful to know how much of their donations will be spent directly on the organizations' programs, and how much will be spent on other organizational costs, such as overheads and fundraising. Overheads include expenses such as rent and utilities, which must be paid on a regular basis to keep the organization running. Fundraising includes anything the organization does to raise money, such as sending out letters or holding an event which donors pay to attend.

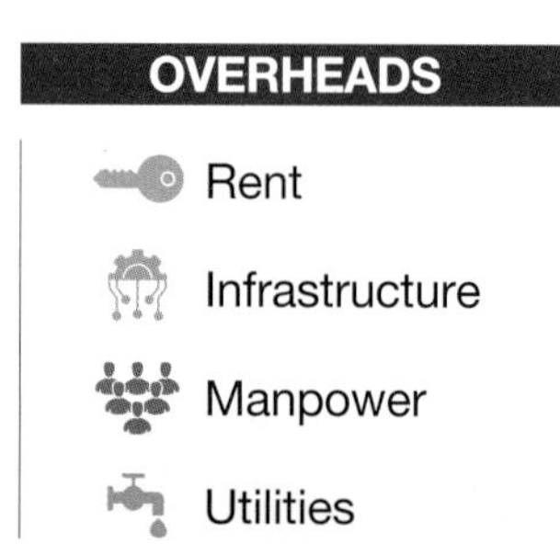


**EXPENDITURE BREAKDOWN**

$100 donation to the **Rainforest Trust**

**Programs:** $92.80 **Fundraising:** $6 **Overheads:** $1.20

$100 donation to **Doctors without Borders**

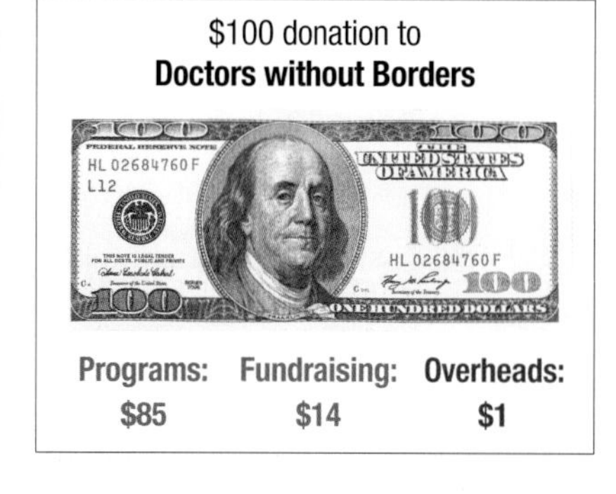

**Programs:** $85 **Fundraising:** $14 **Overheads:** $1

$100 donation to **Helen Keller International**

**Programs:** $83.70 **Fundraising:** $2.80 **Overheads:** $13.50

Sources: Charity Navigator; Doctors Without Borders

**C** Listen to a talk about what you should consider when choosing an organization to donate to. Circle the speaker's main idea.

**a.** When choosing a charity to donate to, consider both its expenses and its total income.

**b.** It's best to donate your money to charitable organizations that use your money carefully.

**c.** You probably shouldn't donate to charitable organizations that have very high expenses.

**D** Work with a partner. Look at the infographic in Exercise B again. Does it tell you which company uses donations most efficiently? Does it say which company contributes the most in aid?

## COMMUNICATE *Synthesize and evaluate ideas*

**E** Work with a partner. Imagine you had a thousand dollars you wanted to spend on a good cause. Look at the four ways below in which you could divide your money. List the factors you'd consider when deciding how much you'd give to each group.

| | |
|---|---|
| donate cash to a charitable organization | donate cash directly to people in need |
| donate cash to a microcredit bank | donate cash to a local research project |

Factors I'd consider:

**1.** ____________________

**2.** ____________________

**3.** ____________________

**4.** ____________________

**F** Work in a group. Read the quotation. How much do you agree, and why? Does it make sense to support an organization that uses less of the money you give to actually help people?

*"Does it really matter that an organization's expenses are high if it allows them to raise more money for those in need? As the common saying goes, sometimes you have to spend money to make money."*

Runners and walkers begin the 12th Annual EIF Revlon Run/Walk for Women fundraiser in New York, U.S.A.

# Putting It Together

LEARNING OBJECTIVES

- Research, plan, and prepare a persuasive speech on an organization you support
- Use various techniques to make an emotional connection

**ASSIGNMENT**

**Individual presentation:** You are going to give a presentation to convince a group to support an organization you believe in.

## PREPARE

**A** Review the unit. What are some of the charitable organizations mentioned in the unit? What makes them special?

**B** Research online to find out more about a local, national, or international organization that supports a cause you believe in. You could research an organization that you've read about in the unit or find a different one. Use the questions below to guide your research, and note down any useful information.

- Is the organization local, national, or international?
- Who or what does it support?
- Does it encourage the involvement of the local community? How?
- Does it need volunteers? What do volunteers do?
- How much of its income is spent on the cause it supports?
- What kinds of donations does it accept? (cash, goods, etc.) How does it use donations?
- How does it provide aid? (credit, cash, goods, consulting, building infrastructure, etc.)

**C** Plan your presentation. Use the following prompts to help you.

Cause: ______________________________

Organization: ______________________________

Why I care about this cause:

______________________________

______________________________

Why I admire this organization:

______________________________

______________________________

Why my classmates should support it too:

______________________________

______________________________

**D** Look back at the vocabulary, pronunciation, and communication skills you've learned in this unit. What can you use in your presentation? Note any useful language below.

______________________________

______________________________

______________________________

______________________________

**E** Below are some ways to make an emotional connection with your listeners. Think about how you can:

- say something surprising or shocking
- say something inspiring
- tell a moving story

**F** Practice your presentation before you give it. Try to make use of the presentation skill that you've learned.

**Presentation Skill**

**Making an Emotional Connection**

Asha de Vos connects emotionally with her audience by using inspiring words, saying surprising or shocking things, and telling moving stories. Doing this is a good way to get listeners to care more about what you're saying, or even share your point of view.

## PRESENT

**G** Give your presentation to a partner. Watch their presentation and evaluate them using the Presentation Scoring Rubrics in the back of the book.

**H** Discuss your evaluation with your partner. Give feedback on two things they did well and two areas for improvement.

# Checkpoint

Reflect on what you have learned. Check your progress.

**I can ...** ☐ understand and use words related to voluntourism.

| | | | | |
|---|---|---|---|---|
| **authorities** | **conservation** | **dependency** | **equitable** | **expectation** |
| **external** | **guidance** | **inclusive** | **outsider** | **policy** |

☐ understand the antonyms of *external*, *inclusive*, *dependence*, and *outsider*.

☐ watch and understand a talk about the importance of local representation.

☐ use stress inside thought groups.

☐ interpret an infographic about how charitable organizations spend their money.

☐ synthesize and evaluate ideas about various ways to help.

☐ use various techniques to make an emotional connection with listeners

☐ give a persuasive presentation on a cause and organization I support.

Entrepreneur Danielle Baskin, founder and CEO of Inkwell Helmets.

# 6

# Be Your Own Boss

## Q Should we work for ourselves?

In this photo, Danielle Baskin sits with some of her painted bicycle helmets. In 2009, she started designing and painting bicycle helmets for herself and customers. Today, she is the CEO of Inkwell Helmets, a company that turns bike helmets into stylish fashion accessories. Creating a business from the ground up is a dream for many people, but just how achievable is this dream? In this unit, we'll look at some of the things aspiring entrepreneurs should do when starting a business, and the challenges they'll need to overcome.

## THINK and DISCUSS

1 Look at the photo and read the caption. How do you feel about the customized helmets in the photo? Would you consider getting one? Why, or why not?

2 Look at the essential question and the unit introduction. How easy do you think it is to be an entrepreneur?

UNIT 6

A B C D

# Building Vocabulary

LEARNING OBJECTIVES

- Use ten words related to entrepreneurship
- Use the prefix *over-*

## LEARN KEY WORDS

**A** Listen to and read the information in the infographic below. Discuss the questions with a partner.

1. Why do you think corporate employment isn't enough for some people?
2. How is being an entrepreneur different from being an employee?
3. What do you think are some of the "other" reasons in the infographic?

## Starting a Business

For some, corporate employment just isn't enough. Rather than **settle** for a regular job, they choose to go down a different path and start a business of their own. However, becoming an entrepreneur isn't easy. The journey is often long, complex, and uncertain.

While examples of **overnight** success are **striking**, success in business is far from guaranteed. If an entrepreneur is to succeed, it is **crucial** that they are **dedicated**, willing to work long hours, and able to deal with **constantly** changing circumstances and **deadlines**.

In some ways, starting a new business is like placing a high-risk **bet**: there's much to gain, but a lot to lose, too. And even then, success—if it happens at all—usually takes years. Despite this, many people are still drawn to the idea, leaving the safety of their jobs to **pursue** dreams of **wealth** and independence.

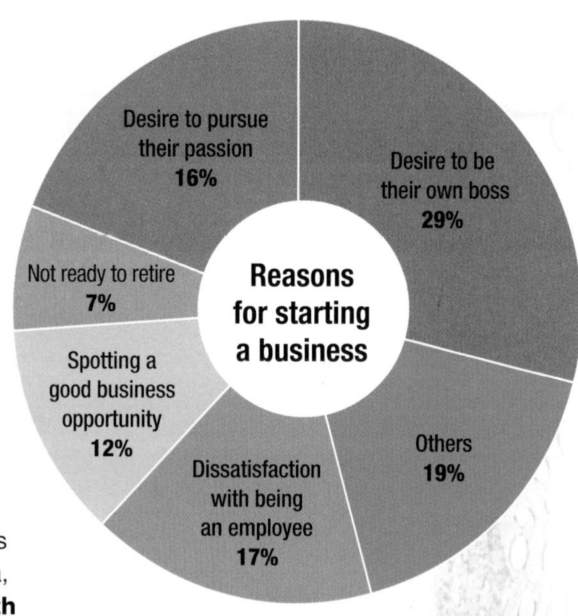


Source: Guidant Financial, Small Business Trends, 2021

**B** Match the correct form of each word in **bold** in Exercise A with its meaning.

1. ________________ all the time, continuously
2. ________________ to accept less than what you really want because it's easier
3. ________________ in a single day, or over a very short period
4. ________________ immediately noticeable, usually in a positive or good sense
5. ________________ a time by which something must be done
6. ________________ a lot of money or things of great value
7. ________________ very committed to doing or achieving something
8. ________________ very important
9. ________________ a choice to risk something in the hope of winning something better
10. ________________ to try to obtain or achieve something, to chase something

**C** The prefix *over-* has several meanings, like *extra, too much,* or *across the span of*. Complete the sentences with the correct answers. Use a dictionary to help you.

| **overall** | **overseas** | **overview** | **overlook** | **overtime** |
|---|---|---|---|---|

1. It's a complex process, so try not to ________________ any of the details.
2. We made a few small mistakes, but ________________, I'm quite pleased with our work.
3. It's a tough job with long hours. She puts in a lot of ________________.
4. As manager, she gets an ________________ of the process that the rest of us don't.
5. They've decided to take their business ________________ to a few foreign markets.

**D** Complete the passage using the correct form of the words in **bold** from Exercise A.

[1] ________________ success in business is rare. In fact, the percentage of businesses that fail is [2] ________________ . 18% fail within a year, 50% fail within five years, and almost 70% fail within ten years. So why do countries support entrepreneurs when new businesses everywhere [3] ________________ struggle just to survive? Entrepreneurship doesn't just help a handful of individuals gain [4] ________________ . It speeds up innovation and creates jobs, which is great for everyone. It is, in fact, [5] ________________ for economic growth.

## COMMUNICATE

**E** Work in a group. Discuss some entrepreneurs you know or have heard about. Were they overnight successes who were just lucky, or did they achieve their success by focusing constantly on their work?

# Viewing and Note-taking

LEARNING OBJECTIVES

- Use abbreviations when taking notes
- Watch a video podcast on tips for entrepreneurs
- Listen to and understand figurative language

## BEFORE VIEWING

**A** Look at the common abbreviations in the box. How are the words abbreviated? What do you notice about the suffixes *-ment* and *-tion*?

| | |
|---|---|
| company → co | equation → eqn |
| department → dept | government → govt |
| English → Eng | people → ppl |
| definition → defn | regarding → re |

**Note-taking Skill**

**Using Abbreviations**

To save time when taking notes, use abbreviated forms of key words. Try abbreviating words by keeping just the first syllable and perhaps a few other useful letters, or by eliminating some or all of the vowels (*a, e, i, o, u*).

**B** Listen to a short talk about trust in businesses. Take notes using abbreviations.

**Example of trust:** Buyg smth online; pay $ bef. rcvg item

- **How trust helps businesses succeed:** new costorbuer feel confitable.
  why do we trust ~~our costober~~ some buj returned buisiness
- **What successful businesses do:** trust the sys using specific tool
  byer trust com
- **Three tools:**
  1. Contract ··· agreements
  2. incentive ··· bonus
  3. transparany ≠ hid info

## WHILE VIEWING

**C** **LISTEN FOR MAIN IDEAS** You are going to watch a video podcast on tips for entrepreneurs. Watch Segment 1 and circle the main purpose below.

**a.** To explain the pros and cons of becoming an entrepreneur

**b.** To explore why some entrepreneurs are more successful than others

**c.** To go over the steps involved in starting your own business

**D** **LISTEN FOR DETAILS** Watch Segment 2 of the video podcast. Complete the notes in the chart below.

**Attitude and personality:**

- *be dedicated and determined*
- *have a ______________ attitude*
- *learn from ______________*
- *be ambitious.*

**Guiding principles:**

1. ______________: *Keep customers ______________, and attract new ones.*
2. ______________: *Hire ______________ and diverse people you enjoy working with.*
3. ______________: *Eat well. ______________ regularly, and get enough ______________.*
4. ______________: *Enjoy the process of building and running your company.*

**E** **LISTEN FOR FIGURATIVE LANGUAGE** Listen to extracts from the podcast where examples of figurative language are used. Match each expression with its meaning.

1. throw in the towel
2. the journey won't be smooth
3. reach greater heights
4. put yourself in someone's shoes

a. achieve more success
b. consider someone's perspective
c. give up, stop
d. progress won't be easy

**Listening Skill**

**Understanding Figurative Language**

It's important to know when speakers are using figurative language. Figurative expressions should not be taken literally. That said, to understand figurative expressions, it helps to visualize what the words are describing.

## AFTER VIEWING

**F** **REFLECT** Are you cut out to be an entrepreneur? Take the quiz below.

1. How much risk are you comfortable with?
   a. None
   b. Some, but only when necessary
   c. A lot. Taking risks is really exciting!
2. How much work are you willing to put in?
   a. A few hours a day
   b. 12 hour days, but not on weekends
   c. As many hours as required
3. How do you prefer to work?
   a. I prefer working independently.
   b. I prefer working with people similar to me.
   c. I like working with different types of people.
4. How well do you deal with failure?
   a. I don't like it. It depresses me.
   b. I don't like it, but I do my best to move on.
   c. Failure is an opportunity to learn.

**G** **EVALUATE** Work in a group. Discuss the questions below.

1. Who do you think is most suited to being an entrepreneur, and why?
2. Share your answers from Exercise F. Do you still feel the same way?

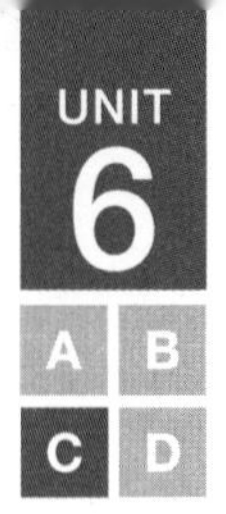


# Noticing Language

LEARNING OBJECTIVES

- Notice language for describing pros and cons
- Discuss pros and cons of a work-related topic

## LISTEN FOR LANGUAGE *Discuss pros and cons*

**A** Read the words and expressions below. Decide if they are positive (**P**), negative (**N**), or could be both (**B**). Discuss with a partner.

| | |
|---|---|
| upside | on the flip side |
| downside | pros |
| the great thing | cons |
| advantage | benefit |
| disadvantage | drawback |

**Communication Skill**

**Discussing Pros and Cons**

When discussing pros and cons, you can use different ways to express the idea of pros and cons, and different adjectives to strengthen your point, for example, *a* ***serious*** *disadvantage, a* ***real*** *drawback, a* ***big*** *plus, a* ***major*** *benefit.*

**B** Listen to the following excerpts from the video podcast in Lesson B. Complete the sentences below with words from Exercise A.

1. Why the popularity? Well, there are a lot of ______ to being an entrepreneur.
2. Another ______ is that entrepreneurs often get to focus on doing what they love.
3. But entrepreneurship has its ______, too. For new businesses, failure is always one small step away.
4. But perhaps the most striking ______ is the fact that entrepreneurs are not guaranteed a stable income.

**C** Work with a partner. Discuss how to complete the sentences below. Then listen and compare your answers.

1. Being a salaried employee has many ______ and ______.
2. You'll have a real ______ over others if you apply early.
3. My job is easy and the hours are great. ______, I don't get paid much.
4. A major ______ of my job is the long drive to work.
5. ______ about this company is that it gets the job done quickly and cheaply.
6. Employees here enjoy many ______, like heath insurance and staff discounts.

According to a 2021 survey of 2,050 U.S. workers, 90% say they are more productive working from home than from the office. Source: Owl Labs, 2021

**D** Complete the conversation below using some words and expressions from the box in Exercise A. Then listen to the conversation and compare your answers. What adjectives could you add to make the pros and cons stronger?

**Anya:** So you've been self-employed for nearly a year now. How's it going?

**Richard:** It's going alright. The [1] ________________ about it is the freedom and flexibility. I get to decide what I spend my time on. But there are some [2] ________________.

**Anya:** Really? Like what?

**Richard:** Honestly, it's pretty stressful—I never really appreciated the [3] ________________ of employment and a stable income until I left my job. And another [4] ________________ is I don't really get weekends anymore. I can't just switch off and forget about work like I used to.

**Anya:** But overall, you're still glad you made the change, aren't you?

**Richard:** Yeah, for now. I think the [5] ________________ outweigh the [6] ________________.

## COMMUNICATE

**E** Choose a topic from the box below. Think of at least three pros and three cons for it.

| | |
|---|---|
| working for a large company | setting up a company with friends |
| having a four-day work week | working from home instead of an office |

**F** Work with a partner. Take turns talking about your topics, discussing the pros and cons of each.

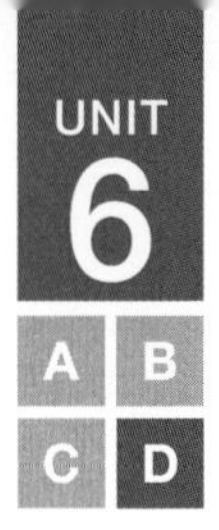


# Communicating Ideas

LEARNING OBJECTIVES

- Use appropriate language for describing pros and cons
- Collaborate to recommend a career choice

**ASSIGNMENT**

**Task:** You are going to collaborate with a partner to learn about an aspiring entrepreneur and recommend a business they could start.

## LISTEN FOR INFORMATION

**A** **LISTEN FOR MAIN IDEAS** Listen to a conversation between two friends, Jodie and Myron. Check (✓) the statements that are true about Jodie.

- ☐ Jodie is going to leave her current job.
- ☐ She is interested in a job that requires working with other people.
- ☐ Getting a job with a stable income is not important to her.
- ☐ She has decided what business she wants to start.

**B** **LISTEN FOR DETAILS** Listen again. Complete the notes below. Use abbreviations where possible.

| Business Idea | Pros | Cons |
|---|---|---|
| do web design | Has the right [1] ___ and experience.<br><br>Has [2] ___ contacts and regular [3] ___. | Wants to do something [4] ___.<br><br>Wants to get away from her desk and work with people more. |
| teach web design | Is good and experienced at web design.<br><br>Would work directly with [5] ___. | Might not guarantee a [6] ___.<br><br>Might not be able to cover her [7] ___. |
| run a coffee shop | Is very passionate about coffee.<br><br>No [8] ___ here, so easier to get regulars. | No café or [9] ___ experience.<br><br>Seems like a big and [10] ___ risk. |

## COLLABORATE

**C** Work with a partner. Discuss the statements in the chart below and check (✓) the jobs that apply. Use the information in Exercise B to help you. When you are done, decide which business idea you think suits Jodie best.

| | **do web design** | **teach web design** | **run a coffee shop** |
|---|---|---|---|
| **Jodie is likely to ...** | | | |
| have the right skills for the job. | | | |
| have the necessary experience for the job. | | | |
| have a strong passion for the job. | | | |
| **The business is likely to ...** | | | |
| have a stable flow of customers. | | | |
| be a safe, low-risk venture. | | | |
| start generating income quicky. | | | |

**D** Work with a new partner. Discuss which new business you think Jodie should start. Use your notes from Exercise C to explain the pros and cons of your recommendation.

# Checkpoint

Reflect on what you have learned. Check your progress.

**I can ...** understand and use words related to entrepreneurship.

| | | | | |
|---|---|---|---|---|
| **bet** | **constantly** | **crucial** | **deadline** | **dedicated** |
| **overnight** | **pursue** | **settle** | **striking** | **wealth** |

use the prefix *over-*.

use abbreviations when taking notes.

watch and understand a video podcast on tips for entrepreneurs.

understand figurative language.

notice language for describing pros and cons.

use language for describing pros and cons to evaluate a career choice.

collaborate and communicate effectively to recommend a career choice.

U.S. charity event company Pallotta TeamWorks created an inspiring new headquarters in a warehouse, with a very low budget.

# Building Vocabulary

LEARNING OBJECTIVES

- Use ten words related to business
- Use antonyms for *tough, infinite, prior,* and *humble*

## LEARN KEY WORDS

**A** Listen to and read the passage below. What two benefits of jobs without titles does the article list? Can you think of other advantages? Discuss with a partner.

**Jobs With No Titles**

The world of business is **tough**. To sustain **revenue** and **growth**, companies must not only hire the best people—they must also retain them. But companies everywhere are all fighting for the same candidates. Therefore, in order to stand out, some choose to employ slightly unusual strategies.

The HR software company Gusto, for example, decided to get rid of job titles. Instead, most of their staff now identify themselves by their teams. Why? Amongst other reasons, the seemingly **infinite** number of job titles in the company was confusing. In fact, the titles were so overly specific that many eligible applicants chose not to try for positions they were qualified for.

So far, Gusto's move has paid off. With job titles gone, team members are now working more closely together. Also, Gusto is attracting employees who would never have applied to them before. While the long-term success of the move is not **guaranteed**, Gusto remains optimistic about its policy.

**B** Work with a partner. Discuss the questions below.

1. Would you like to work for a company without job titles? Why, or why not?
2. What alternative approaches could have solved the problems faced by Gusto?
3. Look at the photo. Would you like to work in an office like this? Why, or why not?

**C** Match the correct form of each word in **bold** from Exercise A with its meaning.

1. growth an increase in size or value
2. infinite limitless, or without end
3. revenue money earned by a business
4. tough difficult and requiring effort to overcome
5. guaranteed promised, or sure to happen

**D** Read the excerpts from Bel Pesce's TED Talk in Lesson F. Choose the options that are closest to the meanings of the words in **bold**.

1. "With that, over three million people downloaded it, over 50,000 people bought physical copies. When I wrote a **sequel**, some impact was guaranteed."

   **a.** the first part of a story **b.** the continuation of a story **c.** a review of a story

2. "I myself have a story in Brazil that people think is an overnight success. I come from a **humble** family ..."

   **a.** inspiring **b.** modest, not rich **c.** unusual

3. "People may think it's an overnight success, but that only worked because for the 17 years **prior** to that, I took life and education seriously."

   **a.** earlier, previous **b.** later **c.** around

4. "It is striking to see how big of an **overlap** there is between the dreams that we have and projects that never happen."

   **a.** a debate or argument **b.** an agreement **c.** a similarity

5. "But OK is never OK. When you're growing towards a **peak**, you need to work harder than ever ..."

   **a.** the end of one's career **b.** the point of greatest success **c.** a major life change

**E** The words in **bold** below are antonyms (opposites). Circle the correct option in each sentence.

1. It was **tough** / **easy** for us to make ends meet during the economic crisis, but things have improved.
2. There are **a limited** / **an infinite** number of positions available, so make sure you show up early.
3. We had a quick chat about the deal and a **prior** / **subsequent** meeting to go over the details.
4. She doesn't act like a rich person, but she's from a very **humble** / **wealthy** family.

## COMMUNICATE

**F** Read the statements. Check (✓) the ones you agree with.

1. ☐ Movie **sequels** are usually better than the originals.
2. ☐ If you work hard, success is **guaranteed**.
3. ☐ Most people reach their **peak** professionally by the age of 35.
4. ☐ Doing something **tough** is more satisfying than doing something easy.
5. ☐ People who come from **humble** backgrounds are more respected when they succeed.

**G** Work with a partner. Look at your answers in Exercise F. Compare and explain your answers.

I disagree with number 1. In my opinion, sequels are usually not as good as the originals.

I think you're right. All the sequels I've seen ...

# Viewing and Note-taking

LEARNING OBJECTIVES

- Watch and understand a talk about ways to achieve our dreams
- Notice the use of intonation and pauses for continuing and concluding



## TEDTALKS

**Bel Pesce**, an entrepreneur and writer, has worked at several big companies including Microsoft, Google, and Deutsche Bank. She has also helped start several businesses. In her TED Talk, *5 Ways to Kill Your Dreams*, she highlights five basic principles that people can follow to transform dreams into reality.

### BEFORE VIEWING

**A** Read the information about Bel Pesce. What is strange about the title of her TED Talk? Discuss with a partner.

**B** Work with a partner. What do you think are the most common mistakes people make when they try to turn a dream into reality?

"Life is never about the goals themselves. Life is about the journey."

## WHILE VIEWING

**C** ▶ **LISTEN FOR DETAILS** Watch Segment 1 of the TED Talk. Complete the notes. Use abbreviations where possible.

TOPIC: How not to ____________

1. Believe in ____________
   - Most of the examples we hear about are ____________
   - A lot of ____________ had to be done first
2. Believe ____________ has the answers you need.
   - People have their own ____________.
   - You still have to ____________ yourself.

**D** ▶ **INFER** Watch Segment 2 of the TED Talk and match quotes 1–4 with the advice a–d.

**1.** ______ "When you're growing toward a peak, you need to work harder than ever and find yourself another peak."

**2.** ______ "But if no one invested in your idea, if no one bought your product, for sure, there is something there that is your fault."

**3.** ______ "... achieving a dream is a momentary sensation, and your life is not. The only way to really achieve all of your dreams is to fully enjoy every step of your journey."

**4.** ______ "Some steps will be right on. Sometimes you will trip. If it's right on, celebrate, because some people wait a lot to celebrate. And if you tripped, turn that into something to learn."

**a.** It is important to both celebrate successes and learn from failures.
**b.** We should not settle for less when we know that more is achievable.
**c.** To succeed, we must enjoy the process, not just achieving our goals.
**d.** When you fail, find out what you did wrong instead of blaming others.

**E** Read the statements below. Check (✓) the ones that Pesce would probably agree with based on this talk.

**1.** ☐ To enjoy overnight success, work extremely hard for a short time.

**2.** ☐ Encourage people with more experience than you to make the most important decisions.

**3.** ☐ When you become successful, set an even higher goal for yourself.

**4.** ☐ Some entrepreneurs fail because they do not take responsibility for their own mistakes.

**5.** ☐ Taking time to celebrate successes will often distract you from your goals.

**6.** ☐ Our goals and the journey we take to achieve them are equally important.

## AFTER VIEWING

**F** **ANALYZE** Read the summary of the things Pesce says NOT to do if you want to follow your dreams. Is each one an attitude (write **A**) or a behavior (write **B**)?

1. Believe in overnight success. ______
2. Rely on someone else to make decisions for you. ______
3. Settle when growth is guaranteed. ______
4. Blame others for your lack of success. ______
5. Focus only on achieving the dream. ______

**G** Work with a partner. Discuss the questions below.

1. Did you find Pesce's unusual way of presenting interesting? Why, or why not?
2. Have you personally had any experiences like those she describes in her talk?
3. Do you agree with her main points? Why, or why not?

## PRONUNCIATION *Continuing and concluding*

**H** Listen to the examples from Pesce's talk below. Mark her pauses with a slash (/). Add arrows to show where she uses rising intonation (↗) and falling intonation (↘).

1. *"And they were going up, and they finally made it to the peak."*
2. *"… and it was a lot of friends, they were going up a mountain, it was a very high mountain, and it was a lot of work."*

**I** Read the four sentences below. Use slashes where would pause and arrows to show where you would use rising or falling intonation.

1. I think that, no matter what the business, success depends not on luck, but on hard work and persistence.
2. The reputation of a business, not how much money it earns, is what is most important.
3. These days, it seems that everyone is starting, or thinking about starting, their own business.
4. A difficult job market, rapid technological change, and overnight success stories make starting my own business sound very appealing.

**Pronunciation Skill**

**Continuing and Concluding**

Speakers often use rising intonation to show that a sentence is continuing (incomplete), and falling intonation when it is concluding (complete). There is also a slight pause after rising intonation, and a longer pause after the falling intonation. This helps listeners follow along, particularly in long sentences with several sections.

**J** Work with a partner. Take turns reading the statements in Exercise I, using appropriate intonation and pausing. Do you agree with the statements? Discuss your ideas.

# Thinking Critically

LEARNING OBJECTIVES

- Interpret an infographic about business failure rates
- Synthesize and evaluate ideas about the reasons for business failures

## ANALYZE INFORMATION

**A** Read the information in the infographic. Use the data to rank the four new business ideas below from 1 (most likely to survive) to 4 (least likely to survive).

1. a business that creates smartphone apps ______
2. a business that provides housing loans ______
3. a business that designs and sells clothes ______
4. a business that trains teachers and tutors ______

**B** Work with a partner. Discuss the questions below.

1. Why do you think new businesses are more likely to succeed in some industries than others?
2. Does a business have to survive for more than ten years before it can be considered successful?

## Which Startups Fail?

Many businesses do not achieve lasting success. While the survival of a business may depend on the kind of industry it's operating in, half of all new businesses in the U.S. are unlikely to survive for more than five years. There are many risks associated with starting a business, and inexperienced business owners may struggle to find their feet.

### New business failure rates

| | |
|---|---|
| By the end of Year 1: | 20% |
| By the end of Year 2: | 30% |
| By the end of Year 5: | 50% |
| By the end of Year 10: | 70% |

### New business failure rates by industry

| | |
|---|---|
| Finance and real estate: | 42% |
| Education, health and agriculture: | 44% |
| Manufacturing: | 51% |
| Construction: | 53% |
| Retail: | 53% |
| Information: | 63% |

Source: National Business Capital, 2019

A shop with few customers in Fall River, U.S.A.

Before the days of online video streaming, DVD rental stores were common.

**C** Listen to a lecture about the reasons why businesses fail and complete the chart below.

| **Company** | Iridium | Blockbuster | Nokia |
|---|---|---|---|
| **Main reason for failure** | Ineffective [1] ______________ | Lack of [3] ______________ | Poor [6] ______________ |
| **Type of company** | Global satellite phone company | [4] ______________ company | Cellphone company |
| **Mistake(s) made** | Built an expensive [2] ______________ that few customers would use | Decided to keep an outdated [5] ______________ | Top and middle management did not work together; lack of focus on [7] ______________ |
| **End result** | Filed for bankruptcy | Filed for bankruptcy | Acquired by Microsoft |

## COMMUNICATE *Synthesize and evaluate ideas*

**D** Work with a partner. Compare the reasons for business failure in Exercise C with the reasons for failing to follow your dreams that Bel Pesce lists in Lesson F. How are they similar? How are they different?

**E** Work with a partner. Look at the different industries listed in the infographic. Will any of the three given reasons for business failure listed in Exercise C affect certain industries more than others? Discuss your ideas.

I think if you're in the retail business, being able to innovate is critical. You need to stand out from all your competitors.

That's true. And retail can also suffer from poor business planning. For example ...

# Putting It Together

LEARNING OBJECTIVES

- Research, plan, and present on a business you would like to start
- Use pausing effectively

**ASSIGNMENT**

**Individual presentation:** You are going to give a presentation to a group of potential investors about a new business that you would like to start.

## PREPARE

**A** Review the unit. What attitudes or personality traits should an aspiring entrepreneur have? What beliefs or principles should they have? Note your ideas. Then check (✓) all the ideas that apply to you.

| Attitude or personality trait | Belief or principle |
|---|---|
| ☐ *Be passionate* | ☐ *Keep attention on target audience* |
| ☐ ________ | ☐ ________ |
| ☐ ________ | ☐ ________ |

**B** Think about what a business should do or avoid doing in order to succeed. Note your ideas. Search online for expert opinions and note down any useful ideas.

**C** Plan your presentation. You are going to talk about your own business idea. Use the chart below to help you.

Business name: ________

Business concept: ________

Why this business is a good idea: ________

________

Possible challenges/limitations: ________

________

Why I would make a good entrepreneur: ________

________

What I can do to help this business succeed: ________

________

**D** Look back at the vocabulary, pronunciation, and communication skills you've learned in this unit. What can you use in your presentation? Note any useful language below.

---

---

---

**E** Below are some ways to practice pausing effectively. Think about how you can:

- write a script for your presentation and annotate it for pausing
- use pauses of different lengths: pause slightly after each thought group, take a breath at each period, and silently count to three after you say anything particularly important.
- practice your presentation with, then without, your script.
- take deep breaths and avoid speaking too quickly.

**Presentation Skill**

**Pausing Effectively**

In Bel Pesce's TED Talk, she uses pauses to help the audience follow her talk. Pauses can be used at different points, such as before a main point is introduced, after a key point is made, and before the speaker moves on to the next main point.

**F** Practice your presentation. Make use of the presentation skill that you've learned.

## PRESENT

**G** Give your presentation to a partner. Watch their presentation and evaluate them using the Presentation Scoring Rubric at the back of the book.

**H** Discuss your evaluation with your partner. Give feedback on two things they did well and two areas for improvement.

# Checkpoint

Reflect on what you have learned. Check your progress.

**I can ...** understand and use words related to business.

| | | | | |
|---|---|---|---|---|
| **growth** | **guaranteed** | **humble** | **infinite** | **overlap** |
| **peak** | **prior** | **revenue** | **sequel** | **tough** |

- use antonyms for tough, *infinite, prior,* and *humble*.
- watch and understand a talk about ways to achieve our dreams.
- notice the use of intonation and pauses for continuing and concluding.
- interpret an infographic about business failure rates.
- synthesize and evaluate ideas about the reasons for business failures.
- use pauses effectively to help listeners process information.
- give a presentation on a business I would like to start.

Visitors to Iceland's Blue Lagoon enjoy bathing in the warm waters.

# 7

# Live Long, Live Well

## What steps can we take to live healthier lives?

This photo shows Iceland's Blue Lagoon which is a popular getaway location for both tourists and locals. Located next to a geothermal power station, its waters are heated by natural volcanic energy. A big part of the lagoon's appeal is the belief that bathing in its warm, mineral-rich water is good for one's health and wellbeing. But while most of us appreciate the importance of healthy living, we don't all have access to hot springs full of healing waters. In this unit, we explore smaller, more practical steps we can all take that could impact our health positively.

## THINK and DISCUSS

1 Look at the photo and read the caption. What positive effects do you think a thermal bath might have for people?

2 Look at the essential question and the unit introduction. What steps can people take to live a healthy life?

UNIT 7

A B C D

# Building Vocabulary

LEARNING OBJECTIVES

- Use ten words related to health
- Understand terms for different systems in our bodies

## LEARN KEY WORDS

**A** Listen to and read the information below. Discuss the questions with a partner.

1. Do you usually get enough sleep?
2. Are you a deep sleeper, or is your sleep usually disrupted?
3. How do you feel when you haven't had enough sleep the night before?

## SLEEP The Simplest Health Hack

Do you often feel sleepy in the afternoon? If you do, there's a good chance you're not getting enough sleep. Why does that matter? According to **numerous** studies, sleep is one of the most important **factors** affecting our health.

Insufficient sleep has been found to have many **unwanted** health effects. It weakens our **immune system**, leads to weight gain, **restricts** our ability to **absorb** new information, and affects our overall brain **function** and health.

Most experts say that we need between seven and nine hours of sleep a night. But it's not just the **duration** of our sleep that matters. We need to minimize sleep **disruptions**, too, if we are to **eliminate** the negative effects of insufficient sleep.

**B** Match the correct form of each word in **bold** in Exercise A with its meaning.

1. ______________ many
2. ______________ something that interferes with a process
3. ______________ to get rid of
4. ______________ the body's mechanisms for dealing with disease and infection
5. ______________ the natural purpose of something
6. ______________ to take something in
7. ______________ something that affects an event, decision, or situation
8. ______________ not desired
9. ______________ to limit something or prevent it from getting too big
10. ______________ the length of time that something lasts

**C** Our bodies have different biological systems. Match them to their functions.

1. immune system ______
2. nervous system ______
3. respiratory system ______
4. digestive system ______
5. cardiovascular system ______

a. circulates blood throughout the body
b. fights against unwanted germs and foreign bodies
c. converts the food we eat to energy and waste
d. transmits signals to and from the brain
e. allows us to breathe and absorb oxygen

**D** Complete the passage. Use the correct form of the **words** in bold from Exercise A.

Sleeping more is perhaps the simplest health hack and the least [1] ______________ to your typical routine. But there are [2] ______________ other ways to boost our health that don't require much effort. Exercise, for example, doesn't have to be exhausting or time consuming. And eating better doesn't mean [3] ______________ all the foods we love from our diets. Small changes to just a few [4] ______________ in our lives can go a long way.

## COMMUNICATE

**E** Discuss with a partner. Why do you think people in different countries get different amounts of sleep? How might the following factors affect sleep duration across countries? Can you think of other factors?

culture economy lifestyle geography

# Viewing and Note-taking

LEARNING OBJECTIVES

- Watch a webinar about forest bathing
- Use outlines to review and organize notes
- Understand unfamiliar words and ideas

## BEFORE VIEWING

**A** Listen to a speaker talking about simple health hacks and complete the notes below. How have the notes been organized?

> **Topic:** Simple Health Hacks
> **Main point 1:** ____________ hacks
> Supporting information:
> 1. Incorporate exercise into daily activities.
>     - take the stairs.
>     - ____________
> 2. Exercise more efficiently.
>     - ____________
>
> **Main point 2:** ____________ hacks
> Supporting information:
> 1. Consume fewer calories.
>     - slow down and create a relaxed mood
>     - ____________

### Note-taking Skill

**Using Outlines to Review and Organize Notes**

It is often useful to reorganize your notes after a lecture, so they are easier to follow when you need to refer to them later. One way of rewriting your notes is by using an outline. An outline can help you understand the connection between ideas, such as the relationship between main ideas and supporting information.

**B** Work with a partner. You are going to watch a webinar about "forest bathing". What do you think it is? What might some of its benefits be?

## WHILE VIEWING

**C** **LISTEN FOR DETAILS** Watch Segment 1 of the webinar and complete the notes below. Were any of your predictions from Exercise B included?

> **Topic:** Forest Bathing
>
> **Main idea 1:** [1] ____________ benefits
> **Supporting info:**
> - stronger [2] ____________
> - reduced blood pressure
> - lower risk of [3] ____________ and diabetes
> - quicker recovery from surgery or illness
> - increased [4] ____________ levels
> - improved [5] ____________
>
> **Main idea 2:** [6] ____________ benefits
> **Supporting info:**
> - lower [7] ____________ levels
> - increased ability to [8] ____________
> - decline in anxiety and depression

**D** **LISTEN FOR MAIN IDEAS** Watch Segment 2 of the webinar. Why does being around trees benefit us? Check (✓) all the reasons that are mentioned.

1. ☐ Humans are genetically conditioned to want to live near trees.
2. ☐ The color green is calming and has a relaxing effect on humans.
3. ☐ Exposure to tiny organisms in forests strengthens our immune systems.
4. ☐ Breathing in chemicals released in forests lowers our stress levels.
5. ☐ We are more alert and focused when surrounded by trees and animals.
6. ☐ Places like forests fill us with wonder and change our perspectives.

**E** **LISTEN FOR DETAILS** Watch Segment 3 of the webinar. Circle **T** (true) or **F** (false).

| | | |
|---|---|---|
| 1. Each session in the woods needs to be at least two hours long. | **T** | **F** |
| 2. You don't have to be very active while forest bathing. | **T** | **F** |
| 3. Tree therapists are people you pay to take you through the forest. | **T** | **F** |
| 4. Forest bathing is an activity that must be done alone. | **T** | **F** |

**F** Work in pairs. Listen to excerpts from the webinar. Write the meanings of the words below.

1. biophilia
_______________
2. phytoncides
_______________
3. solitary
_______________
4. tranquility
_______________

> **Listening Skill**
> **Understanding Unfamiliar Terms**
> When a speaker uses a word or phrase you are unfamiliar with, listen for clues that can help you understand its meaning. For example, listen for synonyms, definitions, explanations, or examples.

## AFTER VIEWING

**G** **EVALUATE** Work with a partner. Answer the questions below.

1. Which of the theories in the webinar did you find convincing? Which were not? Why?
2. Would you try forest bathing? Why, or why not?

UNIT 7

A B C D

# Noticing Language

LEARNING OBJECTIVES

- Notice language used to mark transitions
- Explain a health-related topic to a partner

## LISTEN FOR LANGUAGE *Use signal words and phrases*

**A** Listen to the following excerpts from the webinar in Lesson B. Complete the sentences using signal words and phrases from the box below.

| first, second, ... | one | also | for example |
| --- | --- | --- | --- |
| next | another | in addition | for instance |
| finally | yet another | beyond | such as |

1. We'll be discussing the benefits of surrounding ourselves with trees and spending more time in places ______ forests and parks.
2. ______ explanation involves *biophilia*—a term which comes from the Greek words *bio* (meaning "living things") and *philia* (meaning "a love of").
3. ______ theory centers around stress.
4. ______ improving our immune systems, phytoncides lower our production of stress hormones, too, which have been known to have unwanted effects.
5. ______, there's the theory of awe. Michelle Shiota, a professor of social psychology at Arizona State University, explains that ...

### Communication Skill

**Using Signal Words and Phrases to Mark Transitions**

When presenting information, use signal words to mark transitions from one idea to the next. Signal words help make the relationship between ideas clear when you have several main points, supporting ideas, or examples to list.

**B** Read the summary of the webinar. Circle the correct signal words and phrases. Then listen and check your answers.

There are many physical benefits of forest bathing. [1]**For example / Next / In addition**, spending time with trees strengthens your immune system and lowers blood pressure. There are [2]**another / also / for instance** mental health benefits, [3]**in addition / for instance / next**, stress reduction and a decline in anxiety and depression. [4]**Another / Finally / Next** mental health benefit is an increase in the ability to focus.

90-year-old Hoei Tobaru works in his garden on the small island of Taketomi, Japan.

**C** Listen to a short lecture on blue-zone communities. Take notes to complete the outline below.

**Blue zones:** places with [1] ______________________________

**Similarities across blue zones:**

**Exercise:**

- Not [2] ______________
- Part of [3] ______________

**Diet:**

- People eat mostly [4] ______________
- Part of culture: diets restricted to food that's [5] ______________
- People don't [6] ______________
  - small [7] ______________
  - single [8] ______________

**Social Networks:**

- People live in the [9] ______________ for generations.
- People have the same [10] ______________ for their whole lives.
- Social networks are both broad and deep.

**D** Work with a partner. Take turns using your outline from Exercise C to present an oral summary of the lecture. Use signal words and phrases to transition between ideas.

## COMMUNICATE

**E** Choose one of the health-related topics below or think of your own topic. Write notes to prepare to talk about it for one to two minutes. Think of signal words and phrases you could use to transition between ideas.

- The different things somebody you know does to stay healthy and active
- A diet or exercise routine you've heard of that is supposed to work well
- Why some people can stay healthy and active even when they are quite old
- Your own idea: ______________________________

______________________________

______________________________

______________________________

**F** Work with a partner. Take turns talking about your topic. Use signal words and phrases to transition between ideas.

UNIT 7

A B C D

# Communicating Ideas

LEARNING OBJECTIVES

- Use signal words and phrases to mark transitions
- Collaborate to provide and explain recommendations

**ASSIGNMENT**

**Task:** You are going to collaborate with a partner to suggest healthy changes to someone's routines and habits.

## LISTEN FOR INFORMATION

**A** **LISTEN FOR MAIN IDEAS** Listen to Lize talking to her new health and fitness coach. Which statement best describes the situation? Circle the correct answer.

**a.** Lize wants to start a new diet and exercise routine.

**b.** Lize doesn't feel healthy, but doesn't know what to do about it.

**c.** Lize doesn't have enough time to take care of her health.

**B** **LISTEN FOR DETAILS** Listen again to the conversation. Complete the coach's notes about Lize's lifestyle.

**A. Physical activity:** *very active / moderately active / not active*

Daily walking time: approx. ______________________

Misc. notes: ______________________

______________________

**B. Eating habits:** *avg. food intake tbd upon receipt of food log*

Eating pace: *fast / moderate / slow*

Eating position: ______________________

Meal setting: ______________________

Misc. notes: ______________________

______________________

**C. Social life:** *very active / moderately active / not active*

Weekdays: ______________________

Weekends: ______________________

Usual social activities: ______________________

Misc. notes: ______________________

______________________

## COLLABORATE

**C** Work with a partner. Discuss your notes on Lize from Exercise B. Decide on three main areas for improvement and at least two suggestions to help with each area. Take notes in the chart below.

Area for improvement: ______________________________

Suggestion 1: ______________________________

Suggestion 2: ______________________________

Area for improvement: ______________________________

Suggestion 1: ______________________________

Suggestion 2: ______________________________

Area for improvement: ______________________________

Suggestion 1: ______________________________

Suggestion 2: ______________________________

**D** Work with a different partner. Take turns sharing your suggestions from Exercise C and explaining your reasons. Use signal words and phrases to mark transitions.

Lize could definitely include more exercise in her daily life.

Yes, definitely! Maybe she could ...

# Checkpoint

Reflect on what you have learned. Check your progress.

**I can ...** understand and use words related to health.

| | | | | |
|---|---|---|---|---|
| **absorb** | **disruption** | **duration** | **eliminate** | **factor** |
| **function** | **immune system** | **numerous** | **restrict** | **unwanted** |

understand terms for the different systems in our bodies.

watch and understand a webinar about forest bathing.

use outlines to review and organize notes.

understand unfamiliar terms when listening.

notice signal words and phrases used to mark transitions.

explain a health-related topic using signal words and phrases to mark transitions.

collaborate and communicate effectively to provide and explain health-related recommendations.

Audience members rest at the Philharmonie de Paris, in France, during a special musical performance designed to match a full night's sleep.

UNIT 7

E F G H

# Building Vocabulary

LEARNING OBJECTIVES

- Use ten words related to sleep
- Understand and use the prefix *inter-*

## LEARN KEY WORDS

**A** Listen to and read the passage below. How many stages of sleep are there, and what are the characteristics of healthy sleep?

**The Sleep Cycle**

Sleeping is more than just lying down and closing our eyes. It's a complex process that can be **classified** into several separate but **interrelated** stages.

The first stage is simply falling asleep. Our brain activity and breathing slow down, and our eyes and muscles relax. We eventually enter a light sleep, which is the second stage of sleep. This stage typically lasts just a few minutes.

The third stage is called deeper sleep. Our heartbeat slows and our body temperature falls. We spend most of our sleeping time in this stage, which **evidence** suggests is essential for healthy brain function and memory. During the fourth stage—deep sleep—it can be hard to wake someone up. Our heartbeat and breathing rate reach their lowest points, and our brain activity slows down. This stage is crucial for **cell** and tissue repair, and strengthening our immune systems.

Finally, we experience rapid eye movement, or REM. REM sleep begins about an hour and a half after we fall asleep. Our brain waves speed up almost to waking levels, our eyes shift back and forth, and our breathing and heartbeat get faster. This is when people dream.

**Optimal** sleep follows a steady **rhythm**. It is without disruption, and it is also long enough to cycle through all the stages of sleep several times a night.

**B** Work with a partner. Discuss the questions below.

1. Which sleep stage do you think if the most important, and why?
2. Look at the photo of the music performance. Would you attend an all-night event like this? Why, or why not?

**C** Match the correct form of each word in **bold** in Exercise A with its meaning.

1. ______________ the best level or state that can be achieved
2. ______________ the smallest basic part that a plant or animal is made of
3. ______________ to divide into groups with similar qualities or characteristics
4. ______________ a regular series of sounds, movements, or behaviors
5. ______________ connected so that each thing has an impact on the other things
6. ______________ items or information that prove something is true

**D** Read the excerpts from Matt Walker's TED Talk in Lesson F. Choose the options that are closest to the meanings of the words in **bold**.

"And when you put those two groups head to head, what you find is a quite **significant** 40-percent **deficit** in the ability of the brain to make new memories without sleep."

1. **significant**
   **a.** complex and difficult to understand
   **b.** harmful or dangerous in various ways
   **c.** large or important enough to be worth attention

2. **deficit**
   **a.** the condition of not having enough of something
   **b.** the condition of not being important enough
   **c.** the condition of not knowing enough

"And let me just tell you about ... the context of **aging** and dementia. Because it's of course no secret that, as we get older, our learning and memory abilities begin to fade and **decline**."

3. **aging**
   **a.** having a good memory
   **b.** growing old
   **c.** becoming weak

4. **decline**
   **a.** become less
   **b.** disappear
   **c.** forget

**E** The prefix *inter-* means "between." Complete the sentences with the words in the box.

| **international** | **interrelated** | **intersection** | **interact** | **interpersonal** |
|---|---|---|---|---|

1. Mental and physical health are ______________. If one suffers, it's likely that the other will, too.
2. Drivers usually need to slow down when they approach an ______________.
3. People who have trouble with ______________ relationships may feel anxious in social situations.
4. Research suggests that the more you ______________ with friends and family, the healthier you will be.
5. The four countries agreed to draft an ______________ trade deal yesterday.

## COMMUNICATE

**F** Work with a partner. Discuss the questions below. Use the words in **bold** and explain your answers.

1. How do you think sleep researchers collect **evidence** to make their conclusions?
2. Do you think that the **optimal** amount of sleep is the same for everyone?
3. Do you believe that sleep and learning are **interrelated**?
4. Do you think it's possible to cancel out a sleep **deficit** by sleeping more the next day?

How do you think sleep researchers collect evidence?

Maybe they have laboratories where they observe people while they sleep?

# Viewing and Note-taking

LEARNING OBJECTIVES

- Watch and understand a talk about the importance of sleep
- Notice the pronunciation of word endings



## TEDTALKS

**Matt Walker** is a researcher and professor of neuroscience and psychology. He is the founder of the Center for Human Sleep Science at the University of California, Berkeley, and has also written a book called *Why We Sleep*. In his TED Talk *Sleep Is Your Superpower*, Walker talks about the importance of sleep for optimal health and brain function.

### BEFORE VIEWING

**A** Read the information about Matt Walker. Why do you think he calls sleep a "superpower"? Discuss with a partner.

## WHILE VIEWING

**B** ▶ **LISTEN FOR DETAILS** Watch Segment 1 of the TED Talk. Complete the notes below.

**Sleep after learning:**

- Similar to [1] ______________________.
- So that we don't [2] ______________________.

**Sleep before learning:**

- Similar to [3] ______________________.
- So that we can [4] ______________________.

No sleep = 40% deficit in [5] ______________________.

**C** ▶ **LISTEN FOR MAIN IDEAS** Watch Segment 2 of the TED Talk. Check (✓) the correct answers. There may be more than one.

**1.** What is the relationship between sleep and mental health in the elderly?

**a.** ☐ Sleep can help the elderly regain lost memories.
**b.** ☐ Our quality of sleep gets worse as we get older.
**c.** ☐ Sleep disruption and a decline in learning and memory in the elderly are interrelated.
**d.** ☐ Alzheimer's disease often causes people to sleep less.

**2.** What do natural killer cells do?

**a.** ☐ They strengthen our immune system.
**b.** ☐ They find and eliminate things that can make us sick.
**c.** ☐ They kill the weaker cells in our body.
**d.** ☐ They disrupt our sleep cycle.

**D** ▶ **LISTEN FOR DETAILS** Watch Segment 3 of the TED Talk. What are the four sleep tips Walker shares? Take notes, then discuss your answers with a partner.

**1.** ______________________

**2.** ______________________

**3.** ______________________

**4.** ______________________

**WORDS IN THE TALK**

*Swiss Army knife* (n) a small foldable which comes equipped with several other tools
*soapbox rant* (n) an informal term for a speech long speech of complaint given by someone with strong feelings about a topic

## AFTER VIEWING

**E** **EVALUATE** Work with a partner. Consider the statements below. How much do you agree with each of them, and why?

1. Since sleep helps us remember, it's better to study less so that you can sleep more the night before a test.
2. 18 degrees Celsius is the optimal sleep temperature for everyone, no matter where they are from.
3. Night-time work is a health risk even if you are able to get enough sleep regularly during the day.

## PRONUNCIATION *Word endings*

**Pronunciation Skill**

**Pronouncing Word Endings**

Clearly pronouncing word endings will make your English much clearer. A lot of important information comes at the ends of words, such as the *-ed* markers for past tense forms, the *-s* marker for plural forms, and suffixes like *-tion* for a noun, and *-er* for a person. While it is important to clearly pronounce word endings, make sure you stress the correct syllable of the word, even if it's not the final syllable.

**F** Listen to the excerpt below and notice how Matt Walker pronounces the word endings in bold clearly. Then listen to a few more excerpts and circle the words you hear.

*"Let me start with the brain and the function**s** of learn**ing** and memory, because what we've discover**ed** over the past ten or so year**s** is that you need sleep after learning to essential**ly** hit the save button on those new memor**ies** so that you don**'t** forget."*

**Excerpt 1:**

a. individual — individuals
b. assigned — assign
c. experimenter — experimental

**Excerpt 2:**

d. classified — classify
e. probable — probably
f. rhythm — rhythms

**Excerpt 3:**

g. damages — damaging
h. impacts — impact

**G** Work with a partner. Take turns to read out **one** of the words in each pair below. The listener must identify the word they heard. Check each other's pronunciation.

| | | | | | |
|---|---|---|---|---|---|
| 1. factor | factors | 4. classified | classifies | 7. decline | declined |
| 2. significant | significance | 5. eliminate | eliminated | 8. restrict | restricts |
| 3. disruption | disrupted | 6. absorb | absorbs | 9. evidence | evident |

**H** Work with a partner. Read the questions aloud, being careful to pronounce all word endings clearly. Then discuss your answers.

1. Do you go to sleep and wake up on a regular schedule? If not, how does your sleep schedule vary?
2. How much sleep do you get on a typical night? What factors affect how much sleep you get?

# Thinking Critically

**LEARNING OBJECTIVES**

- Interpret an infographic about factors that can contribute to a longer life
- Synthesize and evaluate ideas about various ways to maintain health

## ANALYZE INFORMATION

**A** Look at the infographic and answer the questions below. Then discuss your answers with a partner.

1. Which of the factors are supported by the strongest evidence?
2. Which factors affect men and women in different ways?
3. Which factors relate to your personality more than your actions?
4. What does the infographic suggest about loneliness? Which factors suggest this?

**B** Which of the factors do you find most surprising or interesting? Do you doubt or disagree with anything in the infographic?

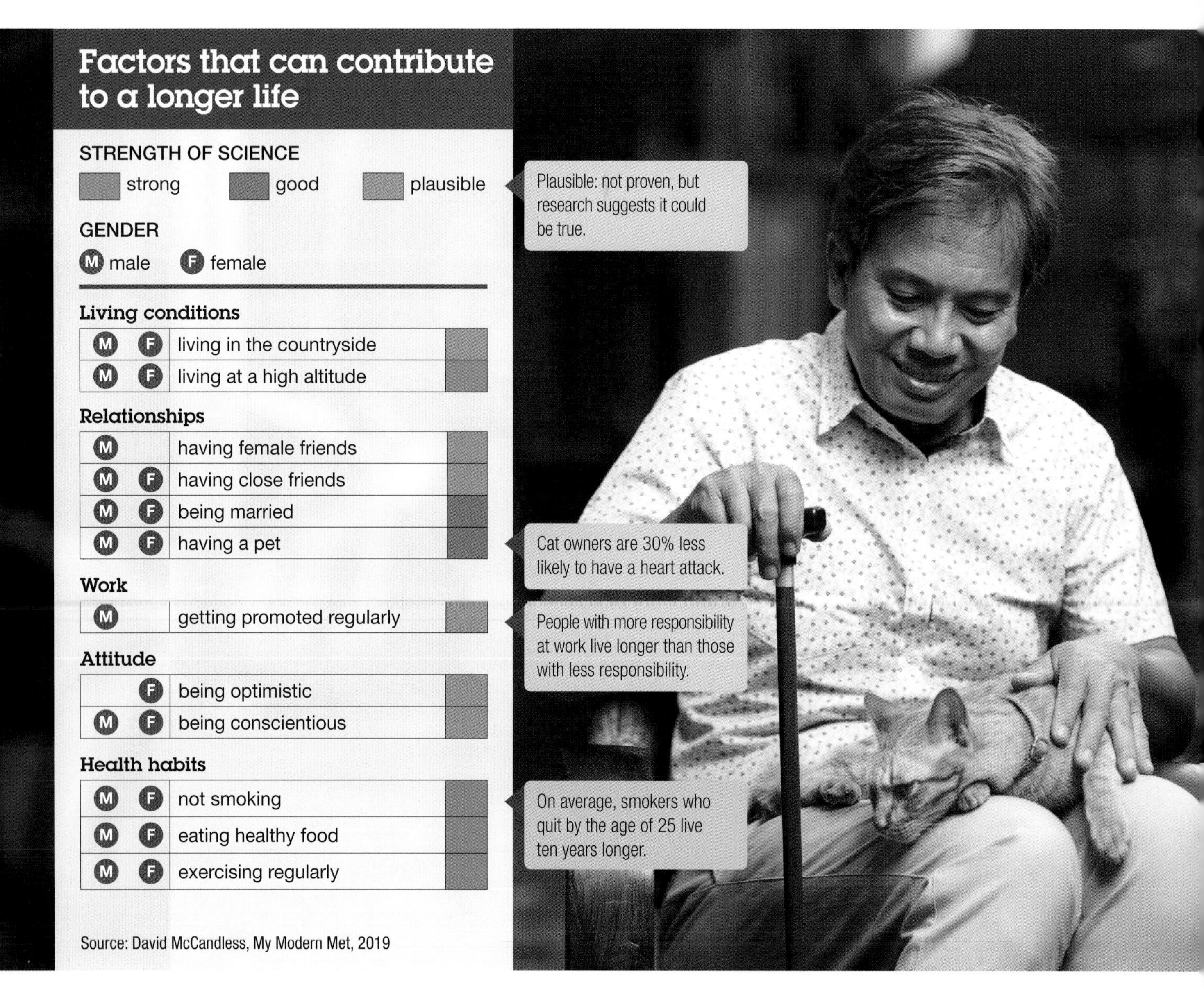

**C** Look at the chart below. Who do you think is most likely to live a long life, according to the infographic? Assume that all three exercise regularly and have healthy diets.

| | **Jose** | **Erik** | **Ling** |
|---|---|---|---|
| **gender:** | male | male | female |
| **lives in:** | the city | the countryside | the mountains |
| **pets:** | yes | no | no |
| **married:** | no | yes | no |
| **attitude:** | optimistic | conscientious | optimistic |
| **smoker:** | no | yes | no |

**D** Work with a partner. On average, women in the U.S. live almost five years longer than men. What might cause this difference in lifespan? Discuss your ideas.

## COMMUNICATE *Synthesize and evaluate ideas*

**E** Consider some of the suggestions for being healthy and living a long life discussed in this unit. Look at the chart and check (✓) your answers.

| **A healthy person ...** | **Do you agree with the statement?** | | **Is this already true for you?** | | **Would you like this to be true for you?** | |
|---|---|---|---|---|---|---|
| | yes | no/unsure | yes | no/rarely | yes | no/unsure |
| spends two hours a week in nature | | | | | | |
| has strong social connections | | | | | | |
| is physically active | | | | | | |
| eats a diet of mostly plants | | | | | | |
| gets 7–9 hours of sleep | | | | | | |
| sleeps at the same time every night | | | | | | |
| is optimistic | | | | | | |
| is married | | | | | | |
| has a pet | | | | | | |
| gets promoted regularly at work | | | | | | |

**F** Work with a partner. Discuss your charts from Exercise E.

1. Which statements do you disagree on, and why?
2. What lifestyle changes is your partner interested in, and why?
3. What tips and recommendations can you offer to your partner?

Maybe you could try to do more social activities, like joining a club.

I think that's a great idea, but what kind of club?

# Putting It Together

LEARNING OBJECTIVES

- Research, plan, and present on a health-related topic
- Organize information in a logical sequence

**ASSIGNMENT**

**Group presentation:** Your group is going to present an argument for or against a statement related to healthy living.

## PREPARE

**A** Review the unit. Which three health tips do you think are supported by the most evidence? Which three are supported by the least evidence?

| Most evidence | Least evidence |
| --- | --- |
| ______________________ | ______________________ |
| ______________________ | ______________________ |
| ______________________ | ______________________ |

**B** Work in groups. Circle the debate topic (1–4) that you think is most interesting. Check (✓) if you are *for* or *against*.

| | for | against |
| --- | --- | --- |
| **1.** Technology is bad for our health. | ☐ | ☐ |
| **2.** Eating meat is good for our health. | ☐ | ☐ |
| **3.** Living alone is bad for our health. | ☐ | ☐ |
| **4.** Forest bathing should be a part of school curriculum. | ☐ | ☐ |

**C** Plan your presentation, based on the topic and position you chose in Exercise B. Anticipate counter-arguments and plan your responses.

| Your side's arguments | Possible counter-arguments | Counter-argument responses |
| --- | --- | --- |
| | | |
| | | |
| | | |
| | | |
| | | |

**D** Look back at the vocabulary, pronunciation, and communication skills you've learned in this unit. What can you use in your presentation? Note any useful language below.

______________________________________________

______________________________________________

______________________________________________

**E** Use the plan below, and your notes from Exercise C, to prepare a detailed outline for your presentation. Make sure you are following a logical sequence.

First speaker:
- Introduce the topic and state your position.
- Give a brief overview of your team's arguments.

Subsequent speakers:
- Present each of your team's arguments in detail.
- Respond to potential counter-arguments.

Final speaker:
- Summarize your team's arguments.
- End with a strong concluding statement.

> **Presentation Skill**
>
> **Organizing Information in a Logical Sequence**
>
> In his TED Talk, Matt Walker presents his ideas in a logical way to make the relationship between ideas easier to understand. To better organize your presentation, try preparing an outline, and think about where you can include signal words. Also, end each main point with a closing statement to make it clear you have finished that point and are moving on to the next one.

**F** Work with your team. Practice your presentation. Make use of the presentation skill that you've learned.

## PRESENT

**G** Give your presentation to another group. Watch their presentation and evaluate them using the Presentation Scoring Rubrics at the back of the book.

**H** Discuss your evaluation with the other group. Give feedback on two things they did well and two areas for improvement.

# Checkpoint

Reflect on what you have learned. Check your progress.

**I can ...**

- [ ] understand and use words related to sleep.

| | | | | |
|---|---|---|---|---|
| **aging** | **cell** | **classify** | **decline** | **deficit** |
| **evidence** | **interrelated** | **optimal** | **rhythm** | **significant** |

- [ ] understand and use the prefix *inter-*.
- [ ] watch and understand a talk about the importance of sleep.
- [ ] notice the pronunciation of word endings.
- [ ] interpret an infographic about factors that can contribute to a longer life.
- [ ] synthesize and evaluate ideas about various ways to maintain health.
- [ ] organize information in a logical sequence.
- [ ] give a presentation arguing for or against a health-related topic.

Jesper Pedersen from Norway competes at the 2022 Winter Paralympics in Yanqing, China.

# 8

# Beyond Limits

## Q How do we define our limits?

We are often told to know our limits, but do limits sometimes prevent us from unlocking our true potential? In the photo, Norwegian skier Jesper Pedersen competes at the 2022 Winter Paralympics in Yanqing, China. His athleticism and skill prove just how far some people can go when they push past the limitations that others see. In this unit, we learn about different limitations people face, and explore how we can transform limitations into opportunities.

## THINK and DISCUSS

1 Look at the photo and read the caption. What is Jesper Pedersen doing? What limitations does he face?

2 Look at the essential question and the unit introduction. What are some limits placed on you by others? What limits do you place on yourself?

# Building Vocabulary

LEARNING OBJECTIVES

- Use ten words related to the brain
- Use forms of *approach, diverse, transform, diagnose,* and *imperfect*

## LEARN KEY WORDS

**A** Listen to and read the information below. What are some factors that affect the health of our brain? Discuss with a partner.

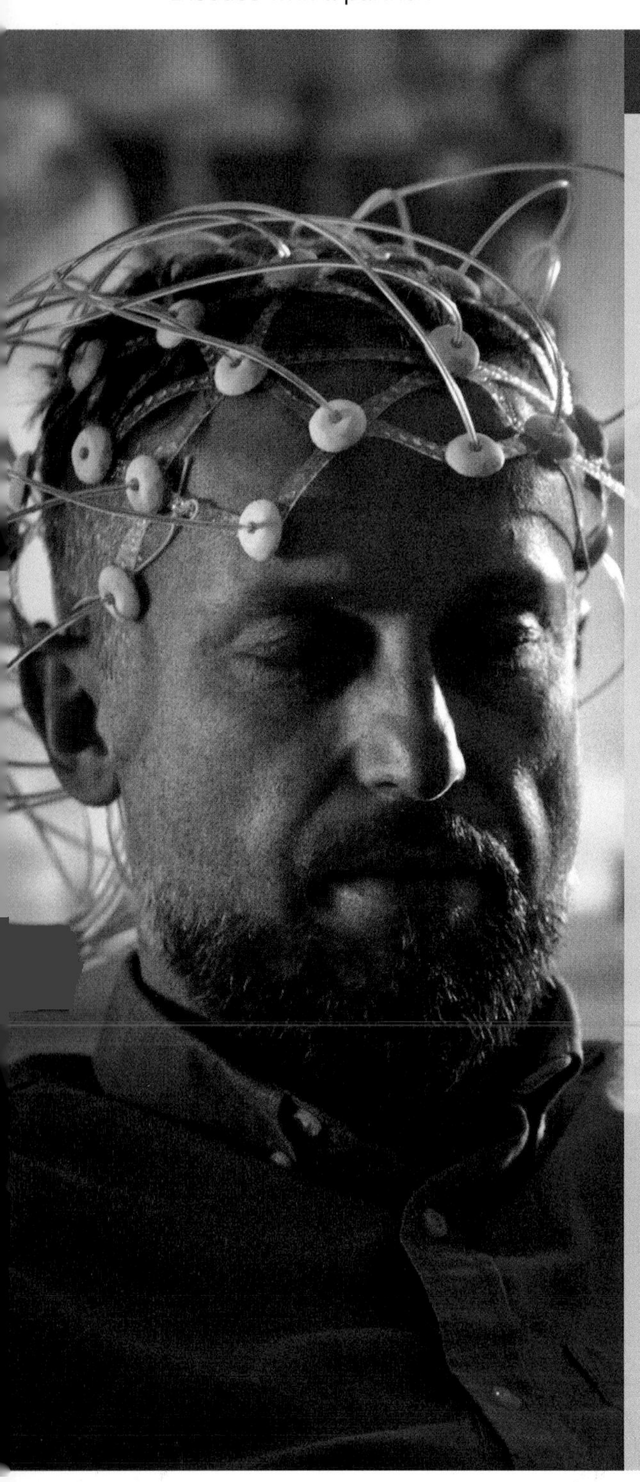

### ALL IN THE BRAIN

The human brain is made up of billions of cells that are linked via trillions of neural connections. It is responsible for our thoughts, actions, and memories. It is also the source of our aspirations and our perceived limitations. It is therefore crucial that we take good care of this precious organ.

Healthy brain function is affected by a **diverse** range of factors. People **diagnosed** with brain injuries, for example, experience not only physical symptoms (such as headaches and dizziness) but mental ones, too (such as memory or concentration problems). Chronic stress can also affect our brains by **eventually** reducing the survival rate of new nerve cells. Staying mentally active, on the other hand, affects our brains positively. There are various ways to do this. For example, one activity **worth** doing is traveling. It broadens our **horizons**, and helps our brains form new connections.

Scientists still have an **imperfect** understanding of how the brain works, but new discoveries are helping to **transform** our **approach** towards treating brain injuries and **disorders**. They are also teaching us more about the vast **spectrum** of factors involved in maintaining a healthy brain.

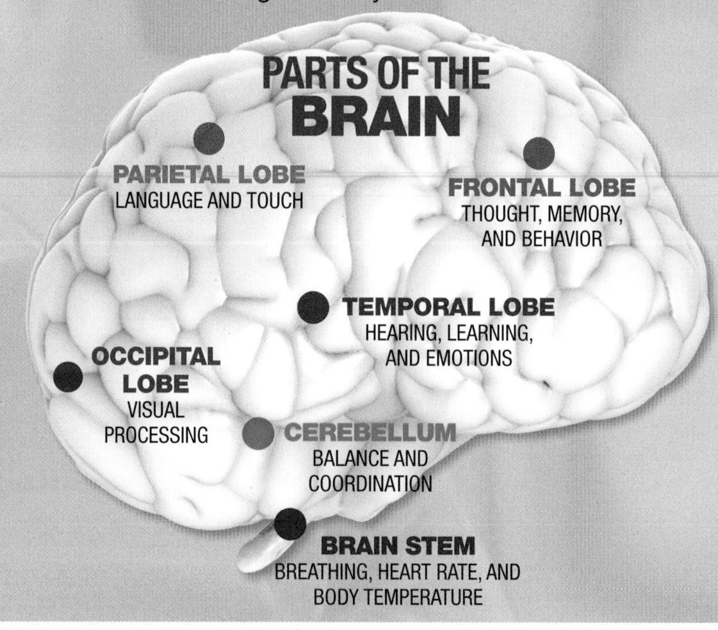

**B** Match the correct form of each word in **bold** in Exercise A with its meaning.

1. ______________ to identify a medical condition
2. ______________ to make a big change in something or someone
3. ______________ the limits of what we know or are interested in
4. ______________ in the end, finally
5. ______________ varied, including many different people or things
6. ______________ a way of doing or thinking about something
7. ______________ having flaws or weaknesses
8. ______________ a range of something
9. ______________ a medical problem or illness
10. ______________ having value

**C** Complete the chart with the correct form of the words. Use a dictionary if necessary.

| Noun | Verb | Adjective |
|---|---|---|
| approach | | X |
| | | diverse |
| | transform | |
| | diagnose | |
| | X | imperfect |

**D** Use the words in the box to complete the statements below.

| **eventually** | **disorder** | **spectrum** | **worth** | **horizons** |
|---|---|---|---|---|

1. Further research will ______________ give us a better understanding of how culture can change the brain.
2. A wide ______________ of problems have developed since the policy was put in place.
3. Exposing yourself to new ideas and different ways of thinking can help you to expand your ______________.
4. Parkinson's disease is a brain ______________ that mainly affects people over 60.
5. It is ______________ spending time and effort looking after your physical and mental health.

## COMMUNICATE

**E** Work with a partner. Look back at the infographic and discuss the questions below.

1. Which functions of the brain do you think are most important? Why?
2. Think of activities that you are able to do well. Which areas of your brain do you think are the most important for those activities?

# Viewing and Note-taking

LEARNING OBJECTIVES

- Shorten common phrases when taking notes
- Watch to a webinar about neurodiversity
- Listen for rhetorical questions

## BEFORE VIEWING

**A** Which phrases do you know how to shorten when taking notes? Which do you think you will use the most often? Discuss with a partner.

| Phrase | Usage / Meaning | Short Form |
|---|---|---|
| *for example / for instance / like* | to give an example | e.g. |
| *in other words / that is* | to introduce a paraphrase or explanation | i.e. |
| *compare with* | to compare one thing with another | cf. |
| *in contrast to / versus* | to contrast one thing with another | vs. |
| *and more / and others / and so on* | to imply that there are other things | etc. |
| *it's important to note that* | to emphasize something | NB |
| *with reference to / with regard to* | to signal you're about to discuss a topic | re: *or* w.r.t. |

**Note-taking Skill**

**Shortening Common Phrases**

When taking notes, many words and phrases get repeated often. Fortunately, there are many commonly used short forms that can replace these words. Short forms like these make your notes shorter and more concise. They also save time, which is important when taking notes while listening.

**B** Listen to a speaker give advice on how to talk about disability. Complete the notes below with some of the shortened forms from the chart above.

**Talking about disabilities**

- The language we use reflects what we think people are capable of.
  It needs to be more sensitive, less limiting.
- Don't say "handicapped," or "disabled person," say "person with disability."
  [1] NB Diff ppl prefer diff terms: ask what they prefer
  (e.g., hearing loss/sight loss [2] vs deaf/blind)
- Remember a so-called disability or disorder is not necessarily a negative.
  [3] e.g., autism is a condition people can live and thrive with.
  Avoid using "disability" or "disorder" if possible.

## WHILE VIEWING

**C** **LISTEN FOR MAIN IDEAS** Listen to Segment 1 of a webinar on neuorodiversity. Check (✓) the ideas that are expressed in the video.

1. ☐ Neurodiversity refers to the range of different ways people think and behave.
2. ☐ Neurodiversity is often associated with conditions such as autism and ADHD.
3. ☐ Neurodivergent people struggle to fit in because they are different.
4. ☐ Neurodivergent people can be taught to think and behave better.
5. ☐ Neurodivergence is a lot more common than many people realize.

**D** **LISTEN FOR DETAILS** Watch Segment 2 of the webinar. Complete the notes below.

- Autism [1] ____________ disorder (ASD): ranges from mild to severe
- Behavior of people with ASD:
  - hard to take part in typical [2] ____________
  - difficulty processing typical [3] ____________ contexts
  - strong and persistent interest in a particular subject
  - like to stick to a [4] ____________
  - show repetitive [5] ____________
- Importance of neurodiversity: ↑ [6] ____________
  - neurodivergent people think differently
  - they have ideas that can benefit everyone (e.g., Einstein, Steve Jobs)
- Dr. Temple Grandin: U.S. cattle expert with autism
  - brain processes [7] ____________ information differently: notices things others miss

**E** Listen to an extract from the webinar. Circle the main reason why the speaker uses a rhetorical question.

**a.** to introduce his central idea

**b.** to signal that details will follow

**c.** to engage his listeners

> **Listening Skill**
>
> **Listening for Rhetorical Questions**
>
> Sometimes speakers ask a question but do not expect an answer. Often, they answer the question themselves. This is a rhetorical question. Listening for rhetorical questions can help you understand what the speaker thinks is important and follow his or her argument more easily.

## AFTER VIEWING

**F** **EVALUATE** Read the quote below from the webinar. To what extent do you agree with the speaker's opinion? Can you think of other situations when having a condition like Dr. Grandin's might be useful?

*"Now, some people may think it must have been very difficult for her to succeed despite that limitation. Well, actually, I believe that Dr. Grandin succeeded, not in spite of her differences or limitations, but because of them."*

# Noticing Language

LEARNING OBJECTIVES

- Notice language used to describe a sequence of events
- Describe a chain of events related to a problem you once encountered

## LISTEN FOR LANGUAGE *Describe a sequence of events*

**A** Read the signal words and phrases below and discuss with a partner. Think about a memorable event that happened to you. Which words and phrases could you use to describe the sequence of events? Are there others that could be added to the list?

**Signal words and phrases to describe a sequence of events:**

| | |
|---|---|
| once | before |
| in (year, month, or season) | as soon as |
| at (time) | when suddenly |
| on (date) | it was then that |
| at that time | eventually |
| after | finally |

**B** Listen to an excerpt from the webinar in Lesson B. What signal words and phrases did the speaker use when describing the sequence of events? Complete the sentences.

**1.** There was ________________ something strange that kept happening at a cattle facility.

**2.** ________________ they contacted Dr. Grandin to see if she could help.

**3.** ________________ Dr. Grandin saw the facility, she identified the problem.

**Communication Skill**

**Describing a Sequence of Events**

When you explain a sequence of events, you can help your listeners understand and follow the order of events by using signal words and phrases.

**C** Listen to a talk about the Indian mountain climber and athlete Arunima Sinha. Then number the events in order from 1 to 6.

**a.** ______ She fought off robbers on a train.

**b.** ______ She played soccer and volleyball.

**c.** ______ She decided to become a mountain climber.

**d.** ______ She successfully climbed the seven highest mountain peaks in the world.

**e.** ______ She lost her left leg.

**f.** ______ She learned to walk with her prosthetic leg.

**D** Work with a partner. Take turns retelling the story from Exercise C. Use signal words and phrases to explain the sequence of events. Did you and your partner choose the same words and phrases?

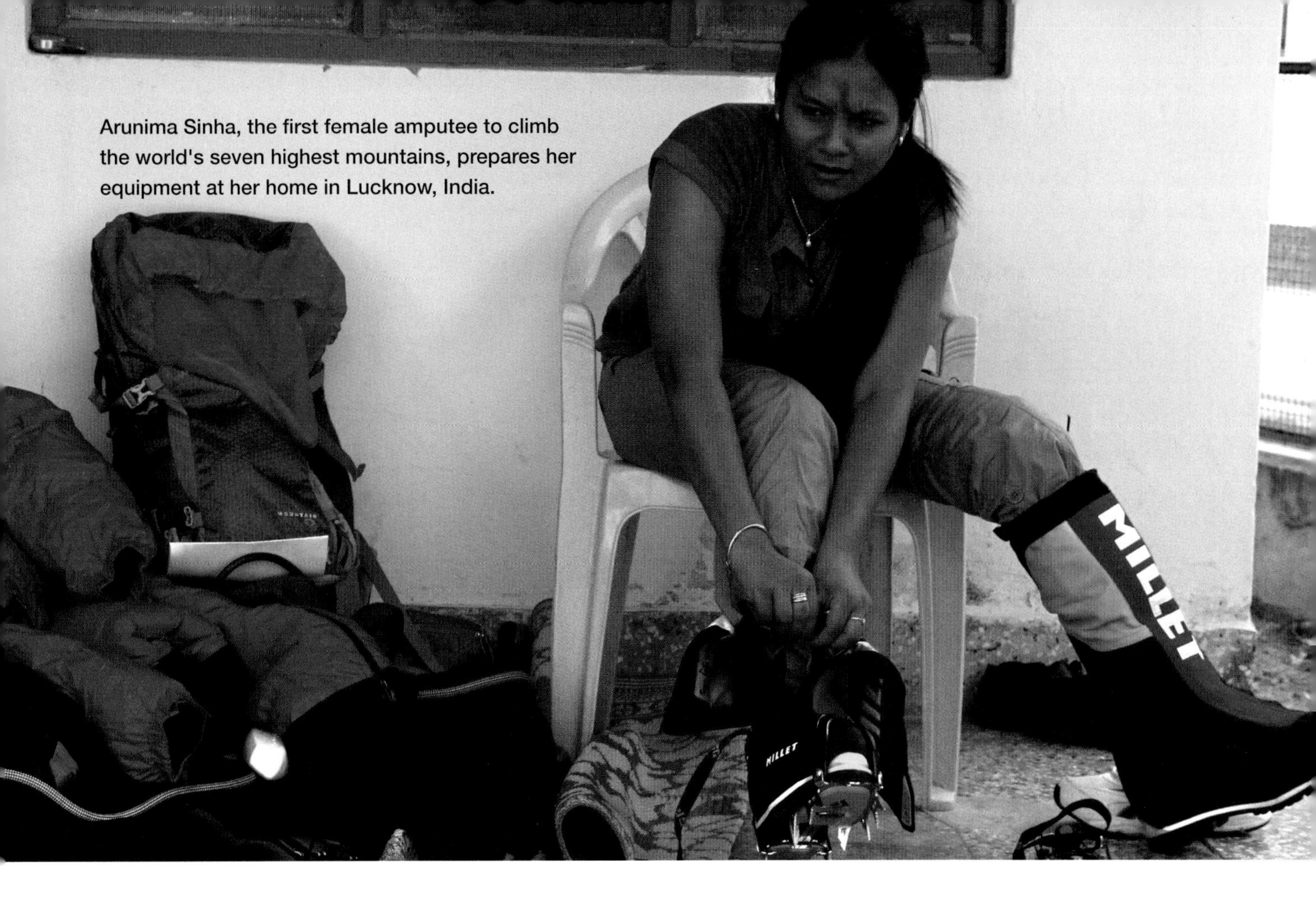


Arunima Sinha, the first female amputee to climb the world's seven highest mountains, prepares her equipment at her home in Lucknow, India.

## COMMUNICATE

**E** Choose one of the contexts from the box below and think about a problem that you once encountered in this context. When did it occur? What did you do about it? What happened after that? Make notes on the sequence of events.

| family | friends | school | work | personal life |
|---|---|---|---|---|

**F** Work in groups. Take turns to share your stories with one another. Use signal words and phrases to help others follow the sequence of events.

> When I was 12, I wanted to go to Disneyland. At that time, my parents promised me that …

# Communicating Ideas

LEARNING OBJECTIVES

- Use appropriate language for explaining problems faced by people with disabilities
- Collaborate to explain how some problems were solved

**ASSIGNMENT**

**Task:** You are going to collaborate in a group to learn about barriers faced by people with disabilities and come up with practical solutions.

## LISTEN FOR INFORMATION

**A** **LISTEN FOR MAIN IDEAS** Listen to a conversation between two students, Julien and Sophie. Which statement below best describes what the conversation is about?

**a.** the different ways people with disabilities access healthcare

**b.** the challenges faced by people with disabilities when accessing healthcare

**c.** the things hospitals do to make visits easier for people with disabilities

**d.** the training healthcare providers undergo to better treat people with disabilities

**B** **LISTEN FOR DETAILS** Listen again and complete the chart.

**Three barriers to healthcare faced by people with disabilities**

Physical:

- *e.g., lack of* [1] ______________ *access in hospitals*

Attitudinal:

- *e.g., providers don't understand how people with different disabilities* [2] ______________

Financial:

- i.e., [3] ______________ *the healthcare they need*

**C** Work with a partner. Look back at your notes in Exercise B. What other barriers might people with disabilities face? Complete the chart with your ideas.

| Physical barriers | Attitudinal barriers | Financial barriers |
| --- | --- | --- |
| | | |

## COLLABORATE

**D** Work in groups. Choose three of the barriers from your notes in Exercise C. Then complete the chart below. Who does it affect, and what is a possible solution to each problem?

| Problem | People affected | Solution |
| --- | --- | --- |
| *E.g. no wheelchair access* | *people with mobility issues* | *install ramps and lifts* |
| | | |
| | | |
| | | |

**E** Work with a partner from a different group. Imagine you are a healthcare worker, and that your partner is a reporter. Explain the problems you identified in Exercise D and the solutions you would like to see.

> What accessibility issues seem common to you?

> One problem is that parking lots and building entrances often aren't wheelchair-friendly ...

# Checkpoint

Reflect on what you have learned. Check your progress.

**I can ...** understand and use words related to the brain.

| | | | | |
| --- | --- | --- | --- | --- |
| **approach** | **diagnose** | **disorder** | **diverse** | **eventually** |
| **horizons** | **imperfect** | **spectrum** | **transform** | **worth** |

understand different forms of *approach, diverse, transform, diagnose,* and *imperfect*.

watch and understand a webinar about neurodiversity.

shorten common phrases when taking notes.

listen for rhetorical questions.

notice language used to describe a sequence of events.

use sequencing language to describe a chain of events related to a problem.

collaborate and communicate effectively to explain some problems facing people with disabilities, and suggest solutions.

Children and teachers from diverse racial and cultural backgrounds enjoy a performance at Millfields Community School, London, U.K.

# Building Vocabulary

LEARNING OBJECTIVES

- Use ten words related to limits
- Use collocations with *encounter*

## LEARN KEY WORDS

**A** Listen to and read the passage below. Who was Jane Elliott and what was her experiment?

### Equality in the classroom

A diverse classroom should contribute to a rich learning environment with opportunities for all. But it is vital that everyone in the classroom is treated fairly and equally.

In the 1960s, an American primary school teacher called Jane Elliott performed a psychology experiment on her students to help them understand this. She informed them that brown-eyed people were better and smarter than blue-eyed people. She then observed their responses.

In just a short while, the brown-eyed children began to **display** signs of aggression toward the blue-eyed children. The discrimination got **progressively** worse throughout the day, and even impacted the students' academic performance. Those in the "superior" group did better in a reading task, while the others **encountered** difficulties accomplishing tasks they had previously found easy. The next week, Elliott repeated the experiment but this time making the blue-eyed children "superior". The results were similar. The students who were considered "superior" **collectively** scored better in tests that day. The others did worse.

Despite its limitations, Elliott's simple experiment became famous across the country. It demonstrated how labeling students could lead to significant changes in academic performance. Though some people were angered by Elliott's methods, her experiment helped us gain a deeper appreciation of the problems created by unjust and **perpetual** discrimination.

**B** Work with a partner. Discuss the questions below.

1. What is discrimination, and how does it relate to Elliott's experiment?
2. Why do you think some people were angered by Elliott's experiment?
3. Look at the photo of the children at a London school. How can students benefit from being in a diverse classroom?

**C** Complete the sentences below using the correct form of the words in **bold** from Exercise A.

1. It's hard not to get upset or angry when you ____________ discrimination.
2. Hearing loss is a problem that ____________ gets worse as we age.
3. The museum is planning to ____________ a giant model of the human brain.
4. The economy seems to be in a state of ____________ decline.
5. It wasn't a solo project. We worked on it ____________.

**D** Read the excerpts from Phil Hansen's TED Talk in Lesson F. Guess the meaning of the words in **bold**.

1. "Now in **hindsight**, it was actually good for some things, like mixing a can of paint or shaking a Polaroid, but at the time this was really doomsday."

   **a.** a prediction of the future   **b.** the assessment of a past situation

2. "The shake developed out of a single-minded **pursuit** ... of making tiny, tiny dots."

   **a.** an attempt to catch or achieve something   **b.** a feeling of intense anger or passion

3. "These dots went from being perfectly round to looking more like tadpoles, because of the shake. So to **compensate**, I'd hold the pen tighter."

   **a.** do something else to make up for something you can't do well   **b.** focus more intensely on a task that's challenging or confusing

4. "So I began experimenting with other ways to **fragment** images where the shake wouldn't affect the work, like dipping my feet in paint and walking on a canvas."

   **a.** make something out of small pieces   **b.** break something up into small pieces

5. "I had this horrible little set of tools, and I felt like I could do so much more with the **supplies** I thought an artist was supposed to have."

   **a.** equipment needed for a job   **b.** people who provide you with equipment

**E** The words in the box all collocate with the verb *encounter*. Complete the sentences using the correct form of the words in the box.

| situation | resistance | obstacle | difficulty |
|---|---|---|---|

1. He thought his idea would be well received, but he encountered ______.
2. If you encounter ______ that involve discrimination, inform your manager.
3. If you take the old mountain trail, you'll probably encounter ______ along the way.
4. The form is short and simple, but let me know if you encounter ______.

## COMMUNICATE

**F** Work with a partner. Discuss the questions below.

1. What goals are you **pursuing** at the moment?
2. What **obstacles** or difficulties have you **encountered** in relation to them?
3. In **hindsight**, should you have done anything differently to better achieve your goals today?

UNIT 8

E F G H

# Viewing and Note-taking

LEARNING OBJECTIVES

- Watch and understand a talk about embracing limitations
- Notice and pronounce *-ed* endings correctly



## TEDTALKS

**Phil Hansen** is a multimedia artist who is known for his unusual artistic methods. In his TED Talk *Embrace the Shake*, he describes a shake in his hand that nearly ended his dream of becoming an artist, and how embracing his condition allowed him to continue making art.

### BEFORE VIEWING

**A** Read the information above about Phil Hansen and his TED Talk. Then read the two meanings of the word *embrace* below. Which meaning applies to the title of the TED Talk? Discuss with a partner.

1. To hold someone in your arms (usually with love or affection)
2. To accept or support something enthusiastically

## WHILE VIEWING

**B** **LISTEN FOR MAIN IDEAS** Watch Segment 1 of Hansen's TED Talk. Which sentence best summarizes what Hansen says?

**a.** We all have limitations that we need to overcome.

**b.** Individual limitations can be opportunities to find creative solutions.

**c.** Having the complete freedom to do whatever you want can be paralyzing.

**d.** To succeed, it's important to place limitations on ourselves.

**C** **SEQUENCE EVENTS** Number the important events in Hansen's life in the correct order. Write *1* next to the first event, *2* next to the second event, and so on. Watch Segment 1 again to check your answers.

**a.** ______ He decided to follow the doctor's advice to "embrace the shake."

**b.** ______ He got a job and was excited to be able to buy lots of art supplies, but he felt paralyzed by all of his choices.

**c.** ______ He decided to become even more creative by placing more limitations on himself.

**d.** ______ He damaged his hand and developed a shake.

**e.** ______ He realized he could still make art if he found different approaches.

**f.** ______ He finally went to see a doctor.

**g.** ______ He quit art school and stopped making art.

**D** **LISTEN FOR DETAILS** Work with a partner. Watch Segment 2 of Hansen's TED Talk. Match the phrases to complete the sentences.

**1.** ______ Hansen wondered if he could become more creative by looking for limitations

**2.** ______ He tried to create art using different tools and techniques,

**3.** ______ He attempted to become an artist without any art,

**4.** ______ By destroying his art, he learned to let go of many things,

**5.** ______ He thinks that by learning to be creative within the confines of limitations,

**a.** e.g., drawing with pencils on Starbucks cups, painting using "karate chops."

**b.** and also found a way to be in a state of constant creation.

**c.** e.g., creating art with less than a dollar's worth of supplies.

**d.** we can transform ourselves and collectively transform the world.

**e.** i.e., create works not to be displayed but to be destroyed.

**WORDS IN THE TALK**

*doomsday* (n) the day on which some people believe the world will end
*pointillism* (n) a form of art where pictures are made with tiny dots
*creative slump* (n) a period where one is unable to think of any new or creative ideas

## AFTER VIEWING

**E** **ANALYZE** Think about the four examples of Hansen's artwork below that you saw in the video. What limitations did Hansen have to overcome for each one? Complete the chart with your notes.

| Pointillist artwork | Paper cup drawing | Canvas hand painting | Matchstick portrait |
|---|---|---|---|
| | | | |

## PRONUNCIATION -ed *endings*

**F** Read the information in the Pronunciation Skill box. Then listen to and read the excerpts below from the TED Talk. Match the underlined words with the categories below.

*"So I dipped my hands in paint, and I just attacked the canvas …"*

*"So I did. I went home, I grabbed a pencil, and I just started letting my hand shake and shake."*

1. Use [d] after vowel sounds and voiced consonants like [n] and [b]. (e.g., *listened, robbed, stayed,* ___ed___)
2. Use [t] after voiceless consonants like [k], [ʃ], and [p]. (e.g., *talked, wished, hoped,* ___t___)
3. Use [ɪd] after the consonant sounds [t] and [d]. (e.g., *shouted, needed,* ___id___)

> **Pronunciation Skill**
>
> **Saying -*ed* Endings**
>
> The word ending -*ed*, which appears in the past tense and past participle forms of many verbs, has three different pronunciations: [d], [t], and [ɪd]. Which sound to use depends largely on the sound before the -*ed*.

**G** Work with a partner. Say the words in the box below. Then write the words in the correct column. Refer to the information in Exercise F for help.

| turned | ended | discovered | developed | worked | limited |
|---|---|---|---|---|---|

| [t] | [d] | [ɪd] |
|---|---|---|
| worked | turned | limited |
| developed | discovered | ended |

**H** Work with a partner. Complete each sentence with a word from the box in Exercise G. Take turns reading the sentences aloud.

1. Unfortunately, the injury ___ended___ her career early.
2. They completed the project despite having had a ___limited___ amount of time and money.
3. All afternoon, the students ___worked___ hard to create the poster.
4. The test was ___developed___ by scientists at a local university.
5. After the break, the team ___turned___ the game around, eventually winning 6–4.

UNIT 8

# Thinking Critically

LEARNING OBJECTIVES

- Interpret an infographic about changing our mindsets
- Synthesize and evaluate ideas about feelings and limitations

## ANALYZE INFORMATION

**A** Look at the infographic below. What is the main idea?

**a.** We should not be too emotional when trying to achieve our goals.
**b.** Feelings sometimes limit us, but it's possible to change how we feel.
**c.** It's important to ignore what others think about us and remain positive.

**B** Work with a partner. Answer the questions below.

**1.** Which step do you think is most difficult when reframing our thoughts?

**2.** How could the student in the passage reframe her situation in a more positive way?

## CHANGING OUR MINDSET

Imagine a student who has been rejected by her first choice of college. How do you think she feels? Like most of us, she'd probably be disappointed or even depressed. But as natural as this may be, do such feelings hold us back?

Feelings affect our thoughts, which in turn limit our actions. We see this all the time. For example, many people who have the ability to achieve their dreams are held back by the fact that they only see their flaws. Reframing is one way to remove such self-imposed limits. By simply changing the way we feel, we can allow ourselves to create a new, more positive reality.

### REFRAMING: A 4-STEP APPROACH

**Step 1**

TAKE ACTION
Consider your current viewpoint.

ASK YOURSELF
How does it make me feel?

**Step 2**

TAKE ACTION
Acknowledge this feeling.

**Step 3**

TAKE ACTION
Look at the story from another angle.

ASK YOURSELF
What other perspectives could I view my situation from?
How does my situation look from these new perspectives?
Does this change how I feel?

**Step 4**

TAKE ACTION
Reframe your situation.
Create a new story.

ASK YOURSELF
How does this story make me feel now?
Do I feel more positive and empowered?

Source: SmartTribes Institute, 2019

**C** Listen to a person talk about how reframing helped him accept who he was. Then read the statements and circle **T** (true) or **F** (false).

**1.** At school, he was only really interested in math. **T** **F**

**2.** He enjoyed socializing with his classmates. **T** **F**

**3.** He needed things to be regular and predictable. **T** **F**

**4.** He saw a doctor because he felt that he wasn't like other people. **T** **F**

**5.** The diagnosis he received made him feel even more different. **T** **F**

**6.** Today, he feels less lonely because he's part of a community. **T** **F**

**D** Work with a partner. Refer to the infographic and apply the four steps of reframing to an aspect of your life that you're dissatisfied with or unhappy about. Then answer the questions below.

**1.** Does reframing help you feel more positive about the situation?

**2.** Is feeling more positive about the situation empowering? Why, or why not?

## COMMUNICATE *Synthesize and evaluate ideas*

**E** Work with a partner. Do the statements below apply mainly to the infographic on reframing or Phil Hansen's advice about embracing our limitations? Write **R** (reframing), **H** (Hansen's advice), or **B** (both).

**1.** Our thoughts can either empower or limit us. ______

**2.** It's important to accept our limitations. ______

**3.** It's important to be aware of our emotions. ______

**4.** It's useful to consider other people's perspectives. ______

**5.** Limitations can sometimes be sources of inspiration. ______

**6.** There's more than one way to think about a problem. ______

**F** Work in groups. Read the opinions A and B below. Which do you agree with more? Why?

**A** I think acknowledging how I feel about a problem isn't really important. What's important is being able to think of solutions. How I feel about it doesn't really change anything.

**B** Maybe it depends on the type of problem you have. If the problem is that you are unhappy about something, then it's very important to acknowledge how you're feeling first. You need to do that before you can start thinking of ways to change things.

# Putting It Together

LEARNING OBJECTIVES

- Research, plan, and present on how someone overcame a limitation
- Use figurative language to make descriptions more vivid

**ASSIGNMENT**

**Individual Presentation:** You are going to give a presentation to a partner about someone who has overcome a limitation to achieve success.

## PREPARE

**A** Review the unit. What are some of the general and specific examples of limitations that were covered? Categorize them in the chart below.

| Physical | Neurological | Social | Psychological |
|---|---|---|---|
| | | | |

**B** Search online for someone inspiring who has overcome a limitation. Use the categories in Exercise A to help you.

**C** Plan your presentation. Use the chart below to help you. Depending on who you choose to talk about, you might need to add, edit, or eliminate questions.

| | |
|---|---|
| **Introduce the person**<br>What is the person's name? How would you describe the person? What limitation did the person face? | |
| **Describe the limitation**<br>What details do you know about it? What difficulties did it cause? What actions did the person take? | |
| **Explain how success was achieved**<br>How did the person make use of their limitation or overcome it? What did you learn from this person? | |

**D** Look back at the vocabulary, pronunciation, and communication skills you've learned in this unit. What can you use in your presentation? Note any useful language below.

______________________________________________

______________________________________________

______________________________________________

**E** Find examples of figurative language in the sentences from Phil Hansen's TED Talk below. Think about how you can use figurative language in your presentation.

*"Now in hindsight, it was actually good for some things, like mixing a can of paint or shaking a Polaroid, but at the time this was really doomsday."*

*"And I was in a dark place for a long time, unable to create."*

**F** Practice your presentation before you give it. Try to make use of the presentation skill that you've learned.

### Presentation Skill

**Using Figurative Language**

In Phil Hansen's TED Talk, he uses figurative language to make his descriptions more vivid. Figurative language such as similes, metaphors, and hyperbole can make your talk come alive. A simile compares one thing with another. A metaphor uses one thing to represent another. And hyperbole refers to statements which are exaggerated for effect.

Simile: The rabbit *ran as quick as lightning* across the path.

Metaphor: Think *outside the box.*

Hyperbole: These shoes are *killing* me!

## PRESENT

**G** Give your presentation to a partner. Watch their presentation and evaluate them using the Presentation Scoring Rubrics at the back of the book.

**H** Discuss your evaluation with your partner. Give feedback on two things they did well and two areas for improvement.

## Checkpoint

Reflect on what you have learned. Check your progress.

**I can ...**

- understand and use words related to limits.

| | | | | |
|---|---|---|---|---|
| **collectively** | **compensate** | **display** | **encounter** | **fragment** |
| **hindsight** | **perpetual** | **progressively** | **pursuit** | **supplies** |

- use collocations with *encounter*.
- watch and understand a talk about embracing limitations.
- pronounce *-ed* endings correctly.
- interpret an infographic about reframing.
- synthesize and evaluate ideas about feelings and limitations.
- use figurative language to make descriptions more vivid.
- give a presentation on how someone overcame a limitation.

# Independent Student Handbook

The Independent Student Handbook is a resource you can use during and after this course. It provides additional support for listening, speaking, note-taking, pronunciation, presentation, and vocabulary skills.

| TABLE OF CONTENTS | PAGE |
|---|---|

# LISTENING AND NOTE-TAKING

## LISTENING STRATEGIES

### Predicting

Speakers giving formal talks usually begin by introducing themselves and then introducing their topic. Listen carefully to the introduction of the topic and try to anticipate what you will hear.

*Strategies:*

- Use visual information including titles on the board, on slides, or in a PowerPoint presentation.
- Think about what you already know about the topic.
- Ask yourself questions that you think the speaker might answer, e.g., *What's the reason for A? How did B happen?*
- Listen for specific introduction phrases (see **Useful Phrases for Presenting**).

### Listening for main ideas

It is often important to be able to tell the difference between a speaker's main ideas and supporting details.

*Strategies:*

- Listen carefully to the introduction. The main idea is often stated at the end of the introduction.
- Listen for rhetorical questions, or questions that the speaker asks and then answers. Often the answer is the main idea.
- Notice ideas that are repeated or rephrased. Repetition and rephrasing often signal main ideas (see **Useful Phrases for Presenting**).

### Listening for details

Supporting details can be a name or a number, an example, or an explanation. When looking for a specific kind of information, it's useful to listen for words that are related to the information you need.

*Strategies:*

- Listen for specific phrases that introduce an example (see **Useful Phrases for Presenting**).
- Notice if an example comes after a general statement from the speaker or is leading into a general statement.
- Notice nouns that might signal causes/reasons (e.g., *factors, influences, causes, reasons*) or effects/results (e.g., *effects, results, outcomes, consequences*).
- Notice verbs that might signal causes/reasons (e.g., *contribute to, affect, influence, determine, produce, result in*) or effects/results (often these are passive, e.g., *is affected by*).
- Listen for specific phrases that introduce reasons/causes and effects/results (see **Useful Phrases for Presenting**).

## Understanding the structure of the presentation

An organized speaker will use certain expressions to alert you to the important information that will follow. Notice signal words and phrases that tell you how the presentation is organized and the relationship between main ideas.

### Introduction

A good introduction includes something like a thesis statement, which identifies the topic and gives an idea of how the lecture or presentation will be organized. Here are some expressions to listen for that indicate a speaker is introducing a topic (see also **Useful Phrases for Presenting**):

*I'll be talking about ...* *My topic is ...*

*There are basically two groups ...* *There are three reasons ...*

### Body

In the body of the lecture, the speaker will usually expand upon the topic. The speaker will use phrases that tell you the order of events or subtopics and their relationship. Here are some expressions to listen for (see also **Useful Phrases for Presenting**):

*The first/next/final (point) is ...* *First/Next/Finally, let's look at ...*

*Another reason is ...* *However, ...*

### Conclusion

In a conclusion, the speaker often summarizes what has been said and may discuss what it means, or make predictions or suggestions. Sometimes speakers ask a question to get the audience to think about the topic. Here are some expressions to listen for (see also **Useful Phrases for Presenting**):

*In conclusion, ...* *In summary, ...*

*As you can see ...* *I/We would recommend ...*

## Understanding meaning from context

Speakers may use words that are new to you, or words that you may not fully understand. In these situations, you can guess the meaning by using the context or situation itself.

*Strategies:*

- Use context clues to guess the meaning of the word, then check if your guess makes sense. What does the speaker say before and after the unfamiliar word? What clues can help you guess the meaning of the word?
- Listen for words and phrases that signal a definition or explanation (see **Useful Phrases for Presenting**).

## Recognizing a speaker's bias

It's important to know if a speaker is objective about the topic. Objective speakers do not express an opinion. Speakers who have a bias or strong feeling about the topic may express views that are subjective.

*Strategies:*

- Notice subjective adjectives, adverbs, and modals that the speaker uses (e.g., *ideal, horribly, should, shouldn't*). These suggest that the speaker has a bias.
- Listen to the speaker's tone. Do they sound excited, happy, or bored?
- When presenting another point of view on the topic, is that other point of view given much less time and attention by the speaker?
- Listen for words that signal opinions (see **Communication Strategies**).

## NOTE-TAKING STRATEGIES

Taking notes is a personalized skill. It is important to develop a note-taking system that works well for you. However, there are some common strategies that you can use to improve your note-taking.

### Before you listen

- Focus. Try to clear your mind before the speaker begins so you can pay attention. If possible, review previous notes or what you already know about the topic.

### As you listen

#### Take notes by hand

Research suggests that taking notes by hand rather than on a laptop or tablet is more effective. Taking notes by hand requires you to summarize, rephrase, and synthesize the information. This helps you *encode* the information, or put it into a form that you can understand and remember.

#### Listen for signal words and phrases

Speakers often use signal words and phrases (see **Useful Phrases for Presenting**) to organize their ideas and indicate what they are going to talk about. Listening for signal words and phrases can help you decide what information to write down in your notes. For example:

*Today we're going to talk about three alternative methods that are ecofriendly, fast, and efficient.*

#### Condense (shorten) information

- As you listen, focus on the most important ideas. The speaker will usually repeat, define, explain, and/or give examples of these ideas. Take notes on these ideas.

    Speaker: *Worldwide, people are using and wasting huge amounts of plastic. For example, Americans throw away 35 million plastic bottles a year.*

    Notes: ***Waste plastic, e.g., U.S. 35 mil plastic bottles/year***

- Don't write full sentences. Write only key words (nouns, verbs, adjectives), phrases, or short sentences.

    Full sentence: *The Maldives built a sea wall around the main island of Malé.*

    Notes: ***Built sea wall—Malé***

- Leave out information that is unnecessary.

    Full sentence: *Van den Bercken fell in love with the music of Handel.*

    Notes: ***VDB loves Handel***

- Write numbers and statistics (***35 mil; 91%***).
- Use abbreviations (***e.g., ft., min., yr***) and symbols (***=, ≠, >, <, %***).
- Use indenting. Write main ideas on the left side of the paper. Indent details.
    - ***Benefits of car sharing***
        - ***Save $***
            - ***Saved $300-400/mo.***
- Write details under key terms to help you remember them.
- Write the definitions of important new words from the presentation.

### After you listen

- Review your notes soon after the lecture or presentation. Add any details you missed.
- Clarify anything you don't understand in your notes with a classmate or teacher.
- Add or highlight main ideas. Cross out details that aren't important or necessary.
- Rewrite anything that is hard to read or understand. Rewrite your notes in an outline or other graphic organizer to record the information more clearly (see **Organizing Information**).
- Use arrows, boxes, diagrams, or other visual cues to show relationships between ideas.

### Organizing information

Sometimes it is helpful to take notes using a graphic organizer. You can use one to take notes while you are listening or to organize your notes after you listen. Here are some examples of graphic organizers:

**Flowcharts** are used to show processes, or cause/effect relationships.

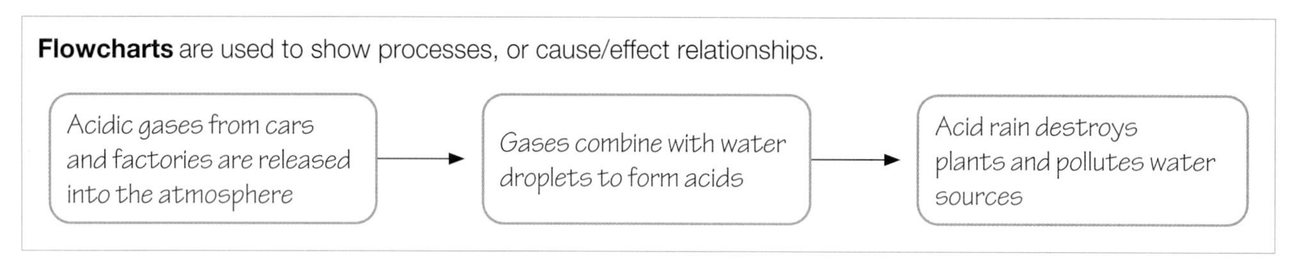


**Mind maps** show the connection between concepts. The main idea is usually in the center with supporting ideas and details around it.

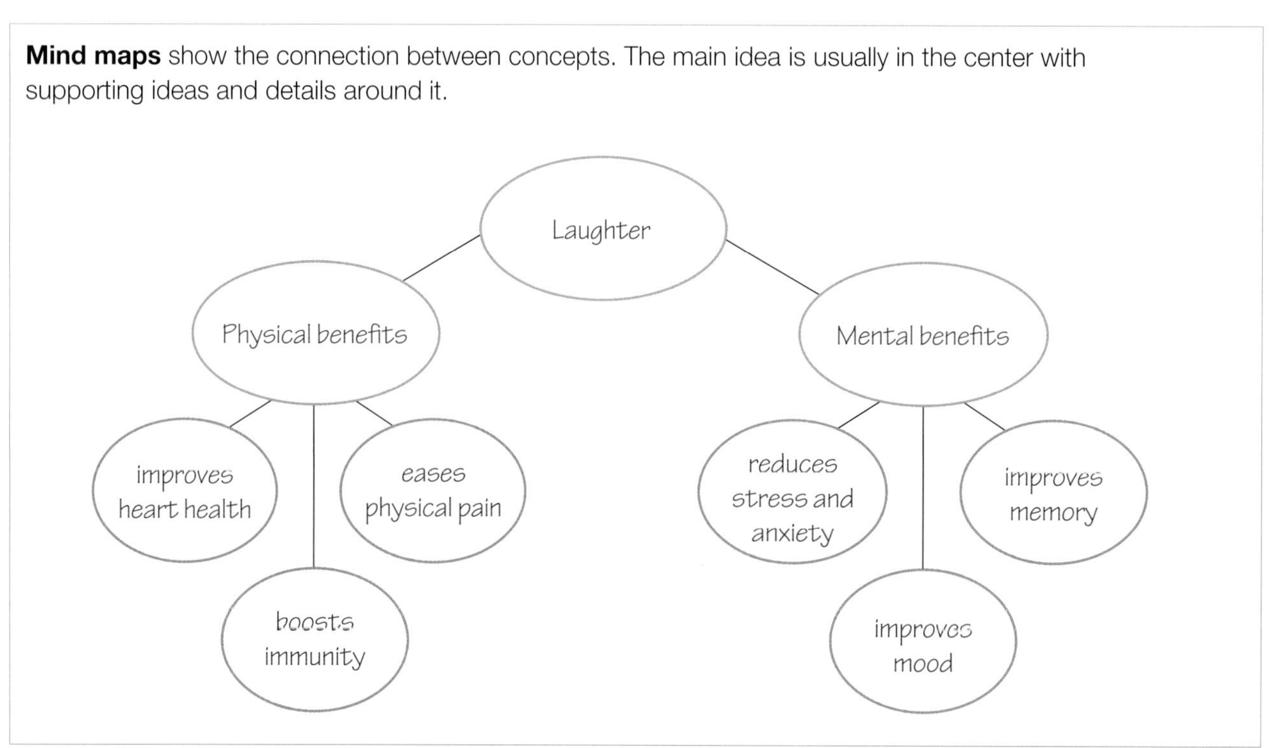

**Outlines** show the relationship between main ideas and details.

To use an outline for taking notes, write the main ideas starting at the left margin of your paper. Below the main ideas, indent and write the supporting ideas and details. You can do this as you listen, or go back and rewrite your notes as an outline later.

1. Saving Water
   A. Why is it crucial to save water?
      i) Save money
      ii) Not enough fresh water in the world

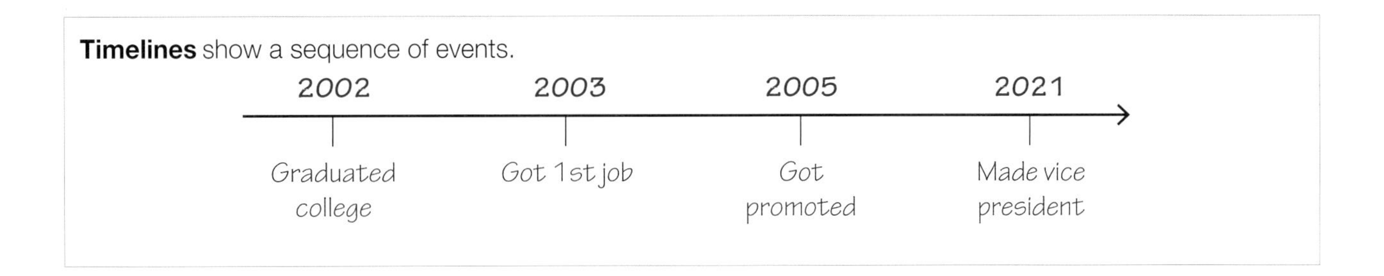

**T-charts** compare two topics.

| Hands-On Learning | |
|---|---|
| **Advantages** | **Disadvantages** |
| 1. Uses all the senses (sight, touch, etc.)<br>2. Encourages student participation<br>3. Helps memory | 1. Requires many types of materials<br>2. May be more difficult to manage large classes<br>3. Requires more teacher time to prepare |

**Timelines** show a sequence of events.

| 2002 | 2003 | 2005 | 2021 |
|---|---|---|---|
| Graduated college | Got 1st job | Got promoted | Made vice president |

**Venn diagrams** compare and contrast two or more topics. The overlapping areas show similarities.

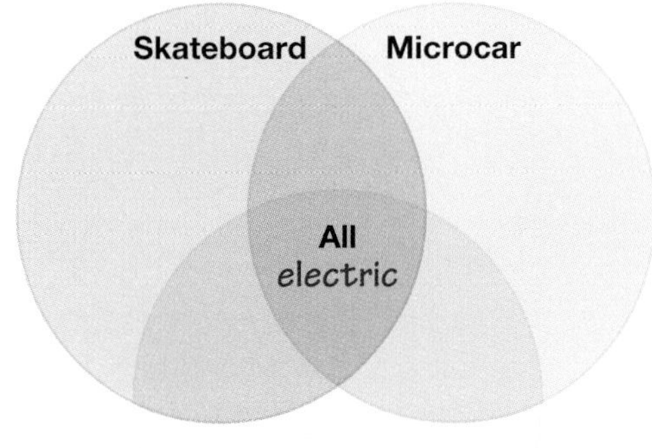

# SPEAKING AND PRONUNCIATION

## COMMUNICATION STRATEGIES

Successful communication requires cooperation from both the listener and speaker. In addition to verbal cues, the speaker can use gestures and other body language to convey their meaning. Similarly, the listener can use a range of verbal and non-verbal cues to show acknowledgment and interest, clarify meaning, and respond appropriately.

## USEFUL PHRASES FOR EXPRESSING YOURSELF

The list below shows some common phrases for expressing ideas and opinions in class.

### Expressing opinions

Your opinion is what you think or feel about something. You can add an adverb or adjective to make your statement stronger.

*I think …*
*I feel …*
*I'm sure …*
*Personally, …*
*If you ask me, …*
*To me, …*
*In my (honest) opinion/view …*
*I strongly believe …*

### Expressing likes and dislikes

There are many expressions you can use to talk about your preferences other than *I like …* and *I don't like…* Using different expressions can help you sound less repetitive.

*I enjoy …*
*I prefer …*
*I love …*
*I don't mind …*
*I can't stand …*
*I hate …*
*I really don't like …*
*I don't care for …*

### Giving facts

Using facts is a good and powerful way to support your ideas and opinions. Your listener will be more likely to believe and trust what you say.

*There is evidence/proof …*
*Experts claim/argue …*
*Studies show …*
*Researchers found …*
*The record shows …*

### Giving tips or suggestions

There are direct and indirect ways of giving suggestions. Imperatives are very direct and let your listener know that it's important they follow your advice. Using questions can make advice sound less direct—it encourages your listener to consider your suggestion.

| Direct | Indirect |
|---|---|
| Imperatives (e.g., *Try to get more sleep.*) | *It's probably a good idea to ...* |
| *You/We should/shouldn't …* | *How about* + (noun/gerund) |
| *You/We ought to …* | *What about* + (noun/gerund) |
| *I suggest (that) ...* | *Why don't we/you …* |
| *Let's …* | *You/We could …* |

## USEFUL PHRASES FOR INTERACTING WITH OTHERS

The list below shows some common phrases for interacting with your classmates during pair and group work exercises.

### Agreeing and disagreeing

In a discussion, you will often need to say whether you agree or disagree with the ideas or opinions shared. It's also good to give reasons for why you agree or disagree.

**Agree**

*I agree.*
*True.*
*Good point.*
*Exactly.*
*Absolutely.*
*Definitely.*
*Right!*
*I was just about to say that.*

**Disagree**

*I disagree.*
*I'm not so sure about that.*
*I don't know.*
*That's a good point, but I don't really agree.*
*I see what you mean, but I think that …*

### Checking your understanding

To make sure that you understand what the speaker has said correctly, sometimes you might need to clarify what you hear. You can check your understanding by rephrasing what the speaker said or by asking for more information.

*Are you saying that … ?*
*So what you mean is … ?*
*What do you mean?*
*How's that?*
*How so?*
*I'm not sure I understand/follow.*
*Do you mean … ?*
*I'm not sure what you mean.*

### Clarifying your meaning

When listeners need to clarify what they hear or understand, speakers need to respond appropriately. Speakers can restate their main points or directly state implied main points.

*What I mean by that is …*
*Not at all.*
*The point I'm making is that …*

### Checking others' understanding

When presenting information that is new to listeners, it's good to ask questions to make sure that your listeners have understood what you said.

*Does that make sense?*
*Do you understand?*
*Do you see what I mean?*
*Is that clear?*
*Are you following me?*
*Do you have any questions?*

### Asking for opinions

When we give an opinion or suggestion, it's good to ask other people for theirs, too. Ask questions to show your desire to hear from your listeners and encourage them to share their views.

*What do you think?*
*Do you have anything to add?*
*What are your thoughts?*
*How do you feel?*
*What's your opinion?*
*We haven't heard from you in a while.*

### Taking turns

During a presentation or discussion, sometimes a listener might want to interrupt the speaker to ask a question or share their opinion. Using questions is a polite way of interrupting. However, the speaker may choose not to allow the interruption, especially if they are about to finish what they have to say.

**Interrupting**

*Excuse me.*
*Pardon me.*
*Can I say something?*
*May I say something?*
*Could I add something?*
*Can I just say ... ?*
*Can I stop you for a second?*
*Sorry to interrupt, but ...*

**Stopping others from interrupting**

*Could I finish what I was saying?*
*If you'd allow me to finish ...*
*Just one more thing.*

**Continuing with your presentation**

*May I continue?*
*Let me finish.*
*Let's get back to ...*

### Asking for repetition

When a speaker speaks too fast or uses words that you are not familiar with, you might want the speaker to repeat themselves. You could apologize first, then politely ask the speaker to repeat what they said.

*Could you say that again?*
*I'm sorry?*
*I didn't catch what you said.*
*I'm sorry. I missed that. What did you say?*
*Could you repeat that, please?*

### Showing interest

It's polite to show interest when you're having a conversation with someone. You can show interest by asking questions or using certain words and phrases. You can also use body language like nodding your head or smiling.

*I see.*
*Good for you.*
*Really?*
*Seriously?*
*Um-hmm.*
*No kidding!*
*Wow.*
*And? (Then what?)*
*That's funny/amazing/incredible/awful!*

## PRONUNCIATION STRATEGIES

When speaking English, it's important to pay attention to the pronunciation of specific sounds. It is also important to learn how to use rhythm, stress, and pausing. Below are some tips about English pronunciation.

### Specific sounds

Research suggests that clear pronunciation of consonant sounds (as compared to vowel sounds) is a lot more useful in helping listeners understand speech. This means that consonant sounds must be accurate for your speech to be clear and easy to understand. For example, /m/ and /n/ are two sounds that sound similar. In a pair of words like *mail* and *nail*, it is important to pronounce the consonant clearly so that the listener knows which word you are referring to.

But there are some exceptions. One example is the pair /ð/ and /θ/, as in *other* and *thing*. These are very often pronounced (both by first and second language English users) as /d/ and /t/ or /v/ and /f/ with little or no impact on intelligibility. There is a lot of variation in vowel sounds in Englishes around the world; however, these differences rarely lead to miscommunication.

| **Vowels** | | | **Consonants** | | |
|---|---|---|---|---|---|
| Symbol | Key Word | Pronunciation | Symbol | Key Word | Pronunciation |
| /ɑ/ | **hot** | /hɑt/ | /b/ | **boy** | /bɔɪ/ |
| | **far** | /fɑr/ | /d/ | **day** | /deɪ/ |
| /æ/ | **cat** | /kæt/ | /ʤ/ | **just** | /ʤʌst/ |
| /aɪ/ | **fine** | /faɪn/ | /f/ | **face** | /feɪs/ |
| /aʊ/ | **house** | /haʊs/ | /g/ | **get** | /gɛt/ |
| /ɛ/ | **bed** | /bɛd/ | /h/ | **hat** | /hæt/ |
| /eɪ/ | **name** | /neɪm/ | /k/ | **car** | /kɑr/ |
| /i/ | **need** | /nid/ | /l/ | **light** | /laɪt/ |
| /ɪ/ | **sit** | /sɪt/ | /m/ | **my** | /maɪ/ |
| /oʊ/ | **go** | /goʊ/ | /n/ | **nine** | /naɪn/ |
| /ʊ/ | **book** | /bʊk/ | /ŋ/ | **sing** | /sɪŋ/ |
| /u/ | **boot** | /but/ | /p/ | **pen** | /pɛn/ |
| /ɔ/ | **dog** | /dɔg/ | /r/ | **right** | /raɪt/ |
| | **four** | /fɔr/ | /s/ | **see** | /si/ |
| /ɔɪ/ | **toy** | /tɔɪ/ | /t/ | **tea** | /ti/ |
| /ʌ/ | **cup** | /kʌp/ | /ʧ/ | **cheap** | /ʧip/ |
| /ɛr/ | **bird** | /bɛrd/ | /v/ | **vote** | /voʊt/ |
| /ə/ | **about** | /ə'baʊt/ | /w/ | **west** | /wɛst/ |
| | **after** | /'æftər/ | /y/ | **yes** | /yɛs/ |
| | | | /z/ | **zoo** | /zu/ |
| | | | /ð/ | **they** | /ðeɪ/ |
| | | | /θ/ | **think** | /θɪŋk/ |
| | | | /ʃ/ | **shoe** | /ʃu/ |
| | | | /ʒ/ | **vision** | /'vɪʒən/ |

Source: *The Newbury House Dictionary plus Grammar Reference*, Fifth Edition, National Geographic Learning/Cengage Learning, 2014.

## Rhythm

The rhythm of English involves stress and pausing.

### Stress

- English words are based on syllables—units of sound that include one vowel sound.
- In every word in English, one syllable has the strongest stress.
- In English, speakers group words that go together based on the meaning and context of the sentence. These groups of words are called *thought groups*. In each thought group, one word is stressed more than the others—the stress is placed on the stressed syllable in this word.
- In general, new ideas and information are stressed.

### Pausing

- Pauses in English can be divided into two groups: long and short pauses.
- English speakers use long pauses to mark the conclusion of a thought, items in a list, or choices given.
- Short pauses are used between thought groups to break up the ideas in sentences into smaller, more manageable chunks of information.

## Intonation

English speakers use intonation, or pitch (the rise and fall of their voice), to help express meaning. For example, speakers usually use a rising intonation at the end of *yes/no* questions, and a falling intonation at the end of *wh-* questions and statements.

# PRESENTING

## PRESENTATION STRATEGIES

The strategies below will help you to prepare, present, and reflect on your presentations.

### Prepare

As you prepare your presentation:

#### Consider your topic

- *Choose a topic you feel passionate about.* If you are passionate about your topic, your audience will be more interested and excited about your topic, too. Focus on one major idea that you can bring to life. The best ideas are the ones your audience wants to experience.

#### Consider your purpose

- *Have a strong beginning*. Use an effective *hook*, such as a quote, an interesting example, a rhetorical question, or a powerful image to get your audience's attention. Include one sentence that explains what you will do in your presentation and why.
- *Stay focused.* Make sure your details and examples support your main points. Avoid sidetracks or unnecessary information that takes you away from your topic.
- *Use visuals that relate to your ideas.* Drawings, photos, video clips, infographics, charts, maps, slides, and physical objects can get your audience's attention and explain ideas effectively, quickly, and clearly. Slides with only key words and phrases can help emphasize your main points. Visuals should be bright, clear, and simple.
- *Have a strong conclusion*. A strong conclusion should serve the same purpose as the strong beginning—to get your audience's attention and make them think. Good conclusions often refer back to the introduction, or beginning, of the presentation. For example, if you ask a question in the beginning, you can answer it in the conclusion. Remember to restate your main points, and add a conclusion device such as a question, a call to action, or a quote.

#### Consider your audience

- *Share a personal story.* You can also present information that will get an emotional reaction; for example, information that will make your audience feel surprised, curious, worried, or upset. This will help your audience relate to you and your topic.
- *Use familiar concepts*. Think about the people in your audience. Ask yourself these questions: Where are they from? How old are they? What is their background? What do they already know about my topic? What information do I need to explain? Use language and concepts they will understand.
- *Be authentic (be yourself).* Write your presentation yourself. Use words that you know and are comfortable using.

#### Rehearse

- *Make an outline.* This will help you organize your ideas.
- *Write notes on notecards*. Do not write full sentences, just key words and phrases to help you remember important ideas. Mark the words you should stress and places to pause.
- *Check the pronunciation of words.* Review the pronunciation skills in your book. For words that you are uncertain about, check with a classmate or a teacher, or look them up in a dictionary. Note and practice the pronunciation of difficult words.
- *Memorize the introduction and conclusion.* Rehearse your presentation several times. Practice saying it out loud to yourself (perhaps in front of a mirror or video recorder) and in front of others.
- *Ask for feedback.* Use feedback and your own performance in rehearsal to help you revise your material. If specific words or phrases are still a problem, rephrase them.

## Present

As you present:

- Pay attention to your pacing (how fast or slow you speak). Remember to speak slowly and clearly. Pause to allow your audience to process information.
- Speak at a volume loud enough to be heard by everyone in the audience, but not too loud. Ask the audience if your volume is OK at the beginning of your talk.
- Vary your intonation. Don't speak in the same tone throughout the talk. Your audience will be more interested if your voice rises and falls, speeds up and slows down to match the ideas you are talking about.
- Be friendly and relaxed with your audience. Remember to smile!
- Show enthusiasm for your topic. Use humor if appropriate.
- Have a relaxed body posture. Don't stand with your arms folded or look down at your notes. Use gestures when helpful to emphasize your points.
- Don't read directly from your notes. Use them to help you remember ideas.
- Don't look at or read from your visuals too much. Use them to support and illustrate your ideas.
- Make frequent eye contact with the entire audience.

## Reflect

As you reflect on your presentation:

- Consider what you think went well during your presentation and what areas you can improve on.
- Get feedback from your classmates and teacher. How do their comments relate to your own thoughts about your presentation? Did they notice things you didn't? How can you use their feedback in your next presentation?

## USEFUL PHRASES FOR PRESENTING

The chart below provides some common signposts and signal words and phrases that speakers use in the introduction, body, and conclusion of a presentation.

| **INTRODUCTION** | |
|---|---|
| **Introducing a topic**<br>*I'm going to talk about …*<br>*My topic is …*<br>*I'm going to present …*<br>*I plan to discuss …*<br>*Let's start with …*<br>*Today we're going to talk about …* | *So we're going to show you …*<br>*Now/Right/So/Well,* (pause) *let's look at …*<br>*There are three groups/reasons/effects/factors …*<br>*There are four steps in this process.* |
| **BODY** | |
| **Listing or sequencing**<br>*First/First of all/The first* (noun)*/To start/To begin, …*<br>*Second/Secondly/The second/Next/Another/Also/ Then/In addition, …*<br>*Last/The last/Finally …*<br>*There are many/several/three types/kinds of/ways, …* | **Signaling problems/solutions**<br>*The problem/issue/challenge (with …) is …*<br>*One solution/answer/response is …* |

| **Giving reasons or causes** | **Giving results or effects** |
|---|---|
| *Because* + (clause): *Because it makes me feel happy …*<br>*Because of* + (noun phrase): *Because of climate change …*<br>*Due to* + (noun phrase) …<br>*Since* + (clause) …<br>*The reason that I like video games is …*<br>*One reason that people do surveys is …*<br>*One factor is* + (noun phrase) …<br>*The main reason that …* | *so* + (clause): *so I decided to try photography*<br>*Therefore,* + (sentence): *Therefore, I changed my diet.*<br>*As a result,* + (sentence).<br>*Consequently,* + (sentence).<br>*… causes* + (noun phrase)<br>*… leads to* + (noun phrase)<br>*… had an impact/effect on* + (noun phrase)<br>*If … then …* |
| **Giving examples** | **Repeating and rephrasing** |
| *The first example is…*<br>*Here's an example of what I mean …*<br>*For instance, …*<br>*For example, …*<br>*Let me give you an example …*<br>*… such as …*<br>*… like …* | *What you need to know is …*<br>*I'll say this again, …*<br>*So again, let me repeat …*<br>*The most important point is …* |
| **Signaling additional examples or ideas** | **Signaling to stop taking notes** |
| *Not only … but*<br>*Besides …*<br>*Not only do … but also* | *You don't need this for the test.*<br>*This information is in your books / on your handout / on the website.*<br>*You don't have to write all this down.* |
| **Identifying a side track** | **Returning to a previous topic** |
| *This is off-topic, …*<br>*On a different subject, …*<br>*As an aside, …*<br>*That reminds me ….* | *Getting back to our previous discussion, …*<br>*To return to our earlier topic …*<br>*OK, getting back on topic …*<br>*So to return to what we were saying, …* |
| **Signaling a definition** | **Talking about visuals** |
| *Which means …*<br>*What that means is …*<br>*Or …*<br>*In other words, …*<br>*Another way to say that is …*<br>*That is …*<br>*That is to say …* | *This graph/infographic/diagram shows/explains …*<br>*The line/box/image represents …*<br>*The main point of this visual is …*<br>*You can see …*<br>*From this we can see …* |
| **CONCLUSION** | |
| **Concluding** | |
| *Well/So, that's how I see it.*<br>*In conclusion, …*<br>*In summary, …*<br>*To sum up, …* | *As you can see, …*<br>*At the end, …*<br>*To review,* + (restatement of main points) |

# BUILDING VOCABULARY

## VOCABULARY LEARNING STRATEGIES

Vocabulary learning is an on-going process. The strategies below will help you learn and remember new vocabulary.

### Guessing meaning from context

You can often guess the meaning of an unfamiliar word by looking at or listening to the words and sentences around it. Speakers usually know when a word is unfamiliar to the audience, or is essential to understanding the main ideas, and will often provide clues to its meaning.

- Restatement or synonyms: A speaker may give a synonym to explain the meaning of a word, using phrases such as *in other words, also called, or ...* , and *also known as*.
- Antonyms: A speaker may define a word by explaining what it is NOT. The speaker might say *unlike A, ...* , or *in contrast to A, B is ...*
- Definitions: Listen for signals such as *which means* or *is defined as*. Definitions can also be signaled by a pause.
- Examples: A speaker may provide examples that can help you figure out what something is. For example, *Paris-Plage is a* ***recreation*** *area on the River Seine, in Paris, France. It has a sandy beach, a swimming pool, and areas for inline skating, playing volleyball, and other activities.*

### Understanding word families: stems, prefixes, and suffixes

Use your understanding of stems, prefixes, and suffixes to recognize unfamiliar words and to expand your vocabulary. A stem is the root part of the word, which provides the main meaning.

A prefix is before the stem and usually modifies meaning (e.g., adding *re-* to a word means "again"). A suffix is after the stem and usually changes the part of speech (e.g., adding *-ation* / *-sion* / *-ion* to a verb changes it to a noun). For example, in the word *endangered*, the stem or root is *danger*, the prefix is *en-*, and the suffix is *-ed*. Words that share the same stem or root belong to the same word family (e.g., *event, eventful, uneventful, uneventfully*).

| Word stem | Meaning | Examples |
|---|---|---|
| *ann* (or *enn*) | year | anniversary, millennium |
| *chron(o)* | time | chronological, synchronize |
| *flex* (or *flect*) | bend | flexible, reflection |
| *graph* | draw, write | graphics, paragraph |
| *lab* | work | labor, collaborate |
| *mob* | move | mobility, automobile |
| *sect* | cut | sector, bisect |
| *vac* | empty | vacant, evacuate |

| Prefix | Meaning | Examples |
|---|---|---|
| *auto-* | self | automatic, autonomy |
| *bi-* | two | bilingual, bicycle |
| *dis-* | not, negation, remove | disappear, disadvantage |
| *inter-* | between | internet, international |
| *mis-* | bad, badly, incorrectly | misunderstand, misjudge |
| *pre-* | before | prehistoric, preheat |
| *re-* | again, back | repeat, return |
| *trans-* | across, beyond | transfer, translate |

| Suffix | Part of speech | Examples |
|---|---|---|
| *-able* (or *-ible*) | adjective | believable, impossible |
| *-en* | verb | lengthen, strengthen |
| *-ful* | adjective | beautiful, successful |
| *-ize* | verb | modernize, summarize |
| *-ly* | adverb; adjective | carefully, happily, friendly, lonely |
| *-ment* | noun | assignment, statement |
| *-tion* (or *-sion*) | noun | education, occasion |
| *-wards* | adverb | backwards, forwards |

### Using a dictionary

A dictionary is a useful tool to help you understand unfamiliar vocabulary you read or hear. Here are some tips for using a dictionary:

- When you see or hear a new word, try to guess its part of speech (noun, verb, adjective, etc.) and meaning, then look it up in a dictionary.
- Some words have multiple meanings. Look up a new word in the dictionary and try to choose the correct meaning for the context. Then see if it makes sense within the context.
- When you look up a word, look at all the definitions to see if there is a basic core meaning. This will help you understand the word when it is used in a different context. Also look at all the related words or words in the same family. This can help you expand your vocabulary. For example, the core meaning of *structure* involves something built or put together.

> **struc·ture** /ˈstrʌktʃər/ *n.* **1** [C] a building of any kind: *A new structure is being built on the corner.* **2** [C] any architectural object of any kind: *The Eiffel Tower is a famous Parisian structure.* **3** [U] the way parts are put together or organized: *the structure of a song||a business's structure*
> —*v.* [T] **-tured, -turing, -tures** to put together or organize parts of s.t.: *We are structuring a plan to hire new teachers.* *-adj.* **structural.**

Source: *The Newbury House Dictionary plus Grammar Reference,* Fifth Edition, National Geographic Learning/Cengage Learning, 2014.

## Multi-word units

You can improve your fluency if you learn and use vocabulary as multi-word units: idioms (*mend fences*), collocations (*trial and error*), and fixed expressions (*in other words*). Some multi-word units can only be understood as a chunk—the individual words do not add up to the same overall meaning. Keep track of multi-word units in a notebook or on note cards.

## Collocations

A collocation is two or more words that often go together. A good way to sound more natural and fluent is to learn and remember as many collocations as you can. Look out for collocations as you read a new text or watch a presentation. Then note them down and try to use them when speaking or in your presentation.

You can organize your notes in a chart to make it easier to review and add to the list as you learn more collocations:

| | |
|---|---|
| share an<br>have an<br>ask for<br>change your | **opinion(s)** |
| fulfill<br>manage<br>set<br>exceed | **expectation(s)** |
| **encounter** | problems<br>resistance<br>obstacles<br>difficulty |

## Vocabulary note cards

You can expand your vocabulary by using vocabulary note cards. Write the word, expression, or sentence that you want to learn on one side. On the other, draw a four-square grid and write the following information in the squares: definition, translation (in your first language), sample sentence, synonyms. Choose words that are high frequency or on the academic word list. If you have looked a word up a few times, you should make a card for it.

| | |
|---|---|
| *definition:* | *first language translation:* |
| *sample sentence:* | *synonyms:* |

Organize the cards in review sets so you can practice them. Don't put words that are similar in spelling or meaning in the same review set, as you may get them mixed up. Go through the cards and test yourself on the meanings of the words or expressions. You can also practice with a partner.

# Presentation Scoring Rubrics

## Unit 1

Presenter(s): ____________________

**The presenter(s) …**

| | Fair 🙂 | Good 😄 | Excellent! 🤩 |
|---|---|---|---|
| presented information in a logical sequence that was easy to follow. | | | |
| spoke clearly with appropriate pacing, volume, and intonation. | | | |
| used humour to connect with an audience. | | | |
| described the success stories of two people. | | | |
| identified useful lessons learned though others' success stories. | | | |
| What did you like? | 1.<br>2. | | |
| What could be improved? | 1.<br>2. | | |

## Unit 2

Presenter(s): ____________________

**The presenter(s) …**

| | Fair 🙂 | Good 😄 | Excellent! 🤩 |
|---|---|---|---|
| presented information in a logical sequence that was easy to follow. | | | |
| spoke clearly with appropriate pacing, volume, and intonation. | | | |
| used enthusiasm to engage the audience. | | | |
| described individuals or groups who are active for a cause. | | | |
| described their actions and impact in the world. | | | |
| What did you like? | 1.<br>2. | | |
| What could be improved? | 1.<br>2. | | |

## Unit 3

Presenter(s): ______________________________

**The presenter(s) ...**

| | Fair | Good | Excellent! |
|---|---|---|---|
| presented information in a logical sequence that was easy to follow. | | | |
| spoke clearly with appropriate pacing, volume, and intonation. | | | |
| encouraged audience participation. | | | |
| identified three uncommon terms and why they are useful or interesting. | | | |
| explained the three new terms clearly, using a range of methods. | | | |
| What did you like? | 1.<br>2. | | |
| What could be improved? | 1.<br>2. | | |

## Unit 4

Presenter(s): ______________________________

**The presenter(s) ...**

| | Fair | Good | Excellent! |
|---|---|---|---|
| presented information in a logical sequence that was easy to follow. | | | |
| spoke clearly with appropriate volume, and intonation. | | | |
| varied the pace to emphasize key points or to create suspense and interest. | | | |
| presented the findings of a survey on stress. | | | |
| drew conclusions from the data and related it to information in this unit. | | | |
| What did you like? | 1.<br>2. | | |
| What could be improved? | 1.<br>2. | | |

## Unit 5

Presenter(s): ______________________________

**The presenter(s) ...**

| | **Fair** 🙂 | **Good** 😄 | **Excellent!** 🤩 |
|---|---|---|---|
| presented information in a logical sequence that was easy to follow. | | | |
| spoke clearly with appropriate pacing, volume, and intonation. | | | |
| used various techniques to make an emotional connection. | | | |
| described an organization that supports a cause. | | | |
| included convincing reasons why others should support this organization. | | | |
| What did you like? | 1.<br>2. | | |
| What could be improved? | 1.<br>2. | | |

## Unit 6

Presenter(s): ______________________________

**The presenter(s) ...**

| | **Fair** 🙂 | **Good** 😄 | **Excellent!** 🤩 |
|---|---|---|---|
| presented information in a logical sequence that was easy to follow. | | | |
| spoke clearly with appropriate pacing, volume, and intonation. | | | |
| used pauses effectively. | | | |
| described and justified a business idea, including some possible challenges. | | | |
| explained why they would be a good person to set up and run this business. | | | |
| What did you like? | 1.<br>2. | | |
| What could be improved? | 1.<br>2. | | |

## Unit 7

Presenter(s): ____________________

**The presenter(s) ...**

| | Fair 🙂 | Good 😄 | Excellent! 🤩 |
|---|---|---|---|
| presented information in a logical sequence that was easy to follow. | | | |
| spoke clearly with appropriate pacing, volume, and intonation. | | | |
| introduced a health-related topic and stated a clear viewpoint on this issue. | | | |
| provided strong arguments supporting the stated viewpoint. | | | |
| anticipated and addressed counter-arguments. | | | |
| What did you like? | 1.<br>2. | | |
| What could be improved? | 1.<br>2. | | |

## Unit 8

Presenter(s): ____________________

**The presenter(s) ...**

| | Fair 🙂 | Good 😄 | Excellent! 🤩 |
|---|---|---|---|
| presented information in a logical sequence that was easy to follow. | | | |
| spoke clearly with appropriate pacing, volume, and intonation. | | | |
| described how a successful person overcame their limitations. | | | |
| used figurative language to make descriptions more vivid. | | | |
| described lessons learned from this person's experiences. | | | |
| What did you like? | 1.<br>2. | | |
| What could be improved? | 1.<br>2. | | |

# Vocabulary Index



**These words are on the Academic Word List (AWL). The AWL is a list of the 570 most frequent word families in academic texts. It does not include the most frequent 2,000 words of English.*

# Credits

**Cover** © Orsolya Haarberg/National Geographic Image Collection; **iii** (from top to bottom, left to right) © Alex McKissack/TED; Nyimas Laula/The New York Times/Redux; © Ryan Lash/TED; © James Duncan Davidson/TED; © Suada Azmy; © Ryan Lash/TED; © Bret Hartman/TED; James Duncan Davidson/TED; **iv** (from top to bottom) © Jared Soares; © Cristina Mittermeier; © Wang Dongling; Ian Forsyth/Getty Images News/Getty Images; Qilai Shen/Panos/Redux; Karsten Moran/Redux; © Guido Cozzi/Atlantide Phototravel; Alexander Hassenstein/Getty Images Sport/Getty Images; **2–3** © Jared Soares; **4** (c) © Frank Juarez; (b) Cengage Learning; **9** RTRO/Alamy Stock Photo; **10** Keith Morris/Alamy Stock Photo; **12** Jamie Squire/Getty Images Sport/Getty Images; **13** Smith Archive/Alamy Stock Photo; **15** © Alex McKissack/TED; **18** (c) Cengage Learning; (bkg) Thomas Barwick/DigitalVision/Getty Images; **22–23** © Cristina Mittermeier; **24** (c) Wavebreakmedia/Shutterstock.com; (br) Cengage Learning; **25** Bettmann/Getty Images; **29** Recep-bg/E+/Getty Images; **32** Kathryn Harms/Dreamstime.com; **35** Nyimas Laula/The New York Times/Redux; **38** Cengage Learning; (bkg) SDI Productions/E+/Getty Images; **42–43** © Wang Dongling; **44** (cr) Philip Scalia/Alamy Stock Photo, (bl) Gas-photo/Shutterstock.com, (br) Heritage Images/Hulton Archive/Getty Images; **45** (cl) GL Archive/Alamy Stock Photo, (bl) J J Osuna Caballero/Alamy Stock Photo; (br) Cengage Learning; **47** Cengage Learning; **49** Myron Standret/Alamy Stock Photo; **52** Xinhua/eyevine/Redux; **55** © Ryan Lash/TED; **57** Kharlanov Evgeny/Shutterstock.com; **58** Cengage Learning; **59** Westend61/Getty Images; **62–63** Ian Forsyth/Getty Images News/Getty Images; **64** Cengage Learning; (bkg)Andy Isaacson/The New York Times/Redux; **69** Steve Debenport/E+/Getty Images; **71** Tao Zhang/Getty Images News/Getty Images; **72** Efrain Padro/Alamy Stock Photo; **75** © James Duncan Davidson/TED; **78** Cengage Learning; **82–83** Qilai Shen/Panos/Redux; **84** (cl) Mark Waugh/Alamy Stock Photo, (b) Christophe Testi/Shutterstock.com; (b) Cengage Learning; **86** Bfk92/E+/Getty Images; **89** Reuters/Alamy Stock Photo; **92** © James P. Blair/National Geographic Image Collection; **95** © Suada Azmy; **98** (cl) Cengage Learning; (b) Anna Kim/iStock/Getty Images; **99** Abaca Press/Alamy Stock Photo; **102–103** Karsten Moran/Redux; **104** (cr) Cengage Learning; (bkg) Alistair Berg/DigitalVision/Getty Images; **109** Thomas Barwick/Stone/Getty Images; **112** © Benny Chan/Fotoworks/Clive Wilkinson Architects; **115** © Ryan Lash/TED; **118** (bkg) Randy Duchaine/Alamy Stock Photo; **119** Henry Iddon/Alamy Stock Photo; **122–123** © Guido Cozzi/Atlantide Phototravel; **124** (bkg) Ricardoreitmeyer/iStock/Getty Images, (br) Christophe Testi/Shutterstock.com; **127** Pakhnyushchy/Shutterstock.com; **128** David McLain/Cavan Images; **130** Roberto Fumagalli/Alamy Stock Photo; **132** © Magnus Wennman/National Geographic Image Collection; **135** © Bret Hartman/TED; **138** Toa55/Shutterstock.com; **142–143** Alexander Hassenstein/Getty Images Sport/Getty Images; **144** (bkg) Janiecbros/E+/Getty Images; (br) Phonlamai Photo/Shutterstock.com; (br) Cengage Learning **149** Strdel/AFP via Getty Images; **152** Gideon Mendel/Corbis Historical/Getty Images; **155** James Duncan Davidson/TED; **158** Damircudic/E+/Getty Images; **162** © Jared Soares, **178–181** stas11/Shutterstock.com

# Acknowledgments

**The authors and publisher would like to thank the following teachers from all over the world for their valuable input during the development process of *21st Century Communication*, Second Edition.**

**Adriana Baiardi**, Colegio Fatima; **Anouchka Rachelson**, Miami Dade College; **Ariya Kilpatrick**, Bellevue College; **Beth Steinbach**, Austin Community College; **Bill Hodges**, University of Guelph; **Carl Vollmer**, Ritsumeikan Uji Junior and Senior High School; **Carol Chan**, National TsingHua University; **Dalit Berkowitz**, Los Angeles City College; **David A. Isaacs**, Hokuriku University; **David Goodman**, National Kaohsiung University of Hospitality and Tourism; **Diana Ord**, Emily Griffith Technical School; **Elizabeth Rodacker**, Bakersfield College; **Emily Brown**, Hillsborough Community College; **Erin Frederickson**, Macomb Community College; **George Rowe**, Bellevue College; **Heba Elhadary**, Gulf University for Science and Technology; **Kaoru Lisa Silverman**, Kyushu Sangyo University; **Lu-Chun Lin**, National Yang Ming Chiao Tung University; **Madison Griffin**, American River College; **Mahmoud Salman**, Global Bilingual Academy; **Marta O. Dmytrenko-Ahrabian**, Wayne State University; **Michael G. Klüg**, Wayne State University; **Monica Courtney**, LaGuardia Community College; **Nora Frisch**, Truckee Meadows Community College; **Pamela Smart-Smith**, Virginia Tech; **Paula González y González**, Colegio Mar del Plata Day School; **Richard Alishio**, North Seattle College; **Rocío Tanzola**, Words; **Shaoyun Ma**, National TsingHua University; **Sorrell Yue**, Fukuoka University; **Susumu Onodera**, Hirosaki University; **Xinyue Hu**, Chongqing No. 2 Foreign Language School; **Yi Shan Tsai**, Golden Apple Language Institute; **Yohei Murayama**, Kagoshima University